Y0-ARU-925

EXPANDING HORIZONS IN BIOETHICS

Expanding Horizons in Bioethics

Edited by

ARTHUR W. GALSTON
Yale University, New Haven, CT, USA

and

CHRISTIANA Z. PEPPARD
Yale University, New Haven, CT, USA

A C.I.P. Catalogue record for this book is available from the Library of Congress.

ISBN 1-4020-3061-4 (HB)
ISBN 1-4020-3062-2 (e-book)

Published by Springer,
P.O. Box 17, 3300 AA Dordrecht, The Netherlands.

Sold and distributed in North, Central and South America
by Springer,
101 Philip Drive, Norwell, MA 02061, U.S.A.

In all other countries, sold and distributed
by Springer,
P.O. Box 322, 3300 AH Dordrecht, The Netherlands.

Printed on acid-free paper

Printed in the Netherlands.

CONTENTS

CONTRIBUTORS

Marcia Angell, M.D., is Senior Lecturer in Social Medicine at Harvard Medical School.

Susannah Baruch, J.D., is a Policy Analyst at the Genetics and Public Policy Center, Berman Bioethics Institute, Johns Hopkins University.

J. Baird Callicott, Ph.D., is Professor of Philosophy, University of North Texas-Denton; and Bioethicist-in-Residence, Interdisciplinary Bioethics Project, Yale University (2004-2005).

Chalmers Clark, Ph.D., is Visiting Scholar in the Donaghue Initiative in Biomedical and Behavioral Research Ethics at the Interdisciplinary Bioethics Project, Yale University (2003-2004).

David Ehrenfeld, M.D., Ph.D., is Professor of Biology at Cook College, Rutgers University.

Ezekiel Emanuel, M.D., Ph.D., is Chair of the Department of Clinical Bioethics, Warren G. Magnuson Clinical Center, National Institutes of Health.

James Flory is a medical student at the University of Pennsylvania and was for two years a Post-Baccalaureate Fellow at the National Institutes of Health.

Arthur Galston, Ph.D., is Emeritus Professor of Biology at Yale University and a founding member of Yale University's Interdisciplinary Bioethics Project.

David Goodstein, Ph.D., is Vice Provost, Professor of Physics and Applied Physics, and Frank J. Gilloon Distinguished Teaching and Service Professor at the California Institute of Technology.

Samuel Gorovitz, Ph.D., is Professor of Philosophy and Professor of Public Administration, Syracuse University; and Bioethicist-in-Residence, Interdisciplinary Bioethics Project, Yale University (2004-2005).

C. Kristina Gunsalus, J.D., is Special Counsel, Adjunct Professor in the College of Law, and Associate Dean of the College of Liberal Arts and Sciences at the University of Illinois at Urbana-Champaign.

Kathy Hudson, Ph.D., is Director of the Genetics and Public Policy Center and Associate Professor in the Berman Bioethics Institute and the Institute of Genetic Medicine, Johns Hopkins University.

Gail Javitt, J.D., M.P.H., is a Policy Analyst at the Genetics and Public Policy Center, Berman Bioethics Institute, Johns Hopkins University; and Adjunct Professor, University of Maryland School of Law.

Karen Lebacqz, Ph.D., is Robert Gordon Sproul Professor of Theological Ethics, Pacific School of Religion.

Ruth Macklin, Ph.D., is Professor of Bioethics at Albert Einstein College of Medicine.

Jonathan Moreno, Ph.D., is Emily Davie and Joseph S. Kornfeld Professor of Biomedical Ethics, and Director, Center for Biomedical Ethics, University of Virginia.

Christiana Peppard is an Editor at Yale University's Interdisciplinary Bioethics Project and a graduate student in Ethics at Yale Divinity School.

Harry M. Rosenberg, Ph.D., was Former Chief (retired), Mortality Statistics Branch, Division of Vital Statistics, National Center for Health Statistics, Centers for Disease Control and Prevention, U.S. Department of Health and Human Services.

George M. Woodwell, Ph.D., is Founder and Director, The Woods Hole Research Center.

ACKNOWLEDGEMENTS

This book is, above all, the result of the willing and enthusiastic work of the contributing authors, many of whom fielded multiple revisions and augmentations to their essays over the course of several years. To the authors, each of whom was a pleasure to have on this project, we owe a great deal of thanks. Our gratitude extends also to the institution that has supported, in various ways, the efforts of both of the editors and the seminar series from which this book originated. Special thanks are due to Donald Green, Director of the Institution for Social and Policy Studies; to Margaret Farley and Robert Levine, Co-Chairpersons of the Interdisciplinary Bioethics Project, whose advice and ongoing support were much appreciated at crucial points during the formulation of this volume; and to Carol Pollard, the indefatigable and inspirational Associate Director of the Interdisciplinary Bioethics Project who first enjoined us to undertake this volume, and who has supported our work every step of the way.

We would also like to recognize the Donaghue Initiative in Biomedical and Behavioral Research Ethics, made possible by the generous support of the Patrick and Catherine Weldon Donaghue Medical Research Foundation, for their support of the Bioethics and Public Policy Seminar Series and the position of Visiting Scholar at the Interdisciplinary Bioethics Project.

A number of individuals have given significant time and service to this project. Helen van der Stelt at Kluwer Academic Publishers has been most helpful throughout the publication process. In addition, this volume benefits enormously from the able and diligent assistance of Zahreen Ghaznavi, Julian Kickbusch, Joanna Kulesz, Anthony Nguyen, Emily O'Neil, and especially Michael Peppard. Finally, special thanks to Lili Beit, who endured many iterations of the volume (and the editors) as *Expanding Horizons in Bioethics* came to fruition.

PREFACE

Like its predecessor, *New Dimensions in Bioethics*, this volume developed out of a series of lectures at Yale University's Institution for Social and Policy Studies. Each speaker in the Bioethics & Public Policy Seminar Series was invited because of her or his expertise in a given area of bioethics. Each of the more successful participants was invited to contribute a manuscript for publication. The essays are bound together by the application of an ethical analysis to scientific questions, and by consideration of policy implications.

At its inception, bioethics was virtually synonymous with medical ethics. As the field grew and attracted new practitioners, it became clear that other applications of this new subject required extension of its scope. For example, environmental ethics, propelled by such authors as Aldo Leopold and Rachel Carson, quickly developed a vigorous literature of its own. More recently, developments in the analysis of the human genome, the enticing medical possibilities offered by the therapeutic use of stem cells, the complexities surrounding the cloning of animals and possibly humans and the development of transgenic agricultural crops have given new impetus to the expansion of traditional bioethical horizons. Bioethics must now adjust to these new realities, for it is clear that public interest in the field is growing as these new challenges appear. This is reflected in the increased rate of book publication, expanded public interest accounts in newspapers, magazines and television and the growing prominence of bioethical discourse in legislatures and governmental policy bodies. Among college students, the interest and excitement generated by bioethics courses can be surprising. For example, a general introduction to bioethics for undergraduates introduced at Yale University two years ago attracted about 350 students in its first year and more than 460 in its second year.

The essays in this book represent topics of exploration in the Bioethics & Public Policy Seminar Series over the past few years. We have divided the topics into three sections: Medical Ethics, Environmental Ethics, and what we have chosen to call

"Science and Society." Together, these categories represent a capsule view of the surprising variety of subjects encompassed within modern bioethics.

Since my formal retirement fourteen years ago from the Department of Molecular, Cellular and Developmental Biology at Yale, I have centered my activities on a remarkably interdisciplinary group at Yale called the Institution for Social and Policy Studies (ISPS). I am grateful to its Director, Professor Donald Green of the Department of Political Science, for encouraging and supporting the various activities in the Interdisciplinary Bioethics Project of ISPS, including the seminar series from which this volume emerged. I owe a great debt to Carol Pollard, Associate Director of the Interdisciplinary Bioethics Project, for facilitating this book and my other work in many ways, and to Professors Margaret Farley and Robert J. Levine, Co-Chairpersons of the Project, for their many suggestions and valuable advice. Finally, it is a pleasure to acknowledge the excellent cooperation and hard work of my Co-Editor, Christiana Peppard. Without her help, this volume could not have shaped up so well.

Arthur W. Galston

New Haven, Connecticut
July 2004

CHRISTIANA PEPPARD

INTRODUCTION

I.

In certain instances, an understanding of what something is not is an instructive prelude to what it is. Such is the case with *Expanding Horizons in Bioethics*: it is a book that fits into the existing literature on bioethics in multifaceted—and, to some degree, anomalous—ways.

Current books within the sprawling scope of bioethics fall into two main categories. Justly respected edited volumes, such as Steinbock, Arras and London's *Ethical Issues in Modern Medicine* or Pojman's *Environmental Ethics,* navigate the terrain of a certain realm of bioethical issues (e.g., medicine or the environment) in as complete a fashion as possible. They delineate foundational principles and challenges to those principles, consider trenchant historical examples, and pose enduring ethical questions. These volumes often serve as fundamental course texts and reference materials. Other types of books on bioethics apply the expertise of one or several primary author(s), and the perspective of a certain discipline, to a particular issue (e.g., physician-assisted suicide), and even to specific circumstances within that issue (e.g., physician-assisted suicide in Oregon after 1997; or the philosophical justifications for assisted suicide).

Expanding Horizons in Bioethics does not fit neatly into either category, yet it incorporates indispensable aspects of both. The book is comprised of fourteen essays divided among three sections: Science and Society, Medical Ethics, and Environmental Ethics. Detailed rationales for each section are below; briefly, we selected the first because the interaction between science and society is increasingly a locus of bioethical inquiry; and we chose the second and third because they constitute significant parts of the bioethical legacy and current debate. Within each essay the tools of a specific discipline are brought to bear in concentrated ways. The fourteen specific issues considered in the three sections of this volume provide a sense of the range of assumptions, methodologies, and goals that permeate current bioethical dialogue.

The goal of this book is to expose the reader to a cross-section of bioethical debates as portrayed by a set of highly qualified scholars, all of whom are well-immersed in their particular fields and familiar with the broader realm of bioethics. *Expanding Horizons in Bioethics* is a compilation that helps the reader explore the variety of questions that inhabit bioethical terrain. Further, the essays serve as examples of some of the many ways in which bioethical evaluation and decision-making may proceed. The essayists were chosen from a series of invited lecturers to

Yale University's Interdisciplinary Bioethics Project between the years 2000 and 2004. Selection was based on the caliber of each scholar's presentation and the methodologies employed in his or her work. We sought variegated approaches to ethical questions, representing a range of academic fields, in order to highlight the interdisciplinary nature of bioethics. Thus, certain premises and methods emerge from each essay that may not be germane to the premises and methods of others. But while the essays are widely roaming, there are also consistent refrains exchanged between seemingly unlikely topics.

These inquiries are not abstractions. The issues are very real; so too are the unresolved questions surrounding them, especially at the level of public policy. Like its predecessor, *New Dimensions in Bioethics* (edited by A. Galston and E. Shurr), this volume portrays a host of quandaries at the intersection of bioethics and public policy. Public policy—understood as the process by which guidelines, regulations, or sanctions are made normative for a society, regardless of whether they originate from the federal branch of government, local government, or no government at all—is an important aspect of bioethical debate and dialogue. At its best, public policy is the result of sustained moral and pragmatic deliberation about an appropriate course of action for the benefit of society. It is often a lens that reflects the norms and values of a certain society. But it is not the only lens.

In this volume, we remain concerned about public policy, but we are fundamentally interested in a larger sense of "policy"—namely, in prudence, wisdom, and the decisional processes of individuals and groups. This entails an eye toward public policy where relevant. Yet we firmly believe that issues such as stem cell research, genetic modification of food sources, end-of-life care, or global climate change are matters that will affect each of our lives—or those of people we know—in significant ways. Therefore, one of our assumptions is that these issues should not be left only to the "experts" or lawmakers; as Ehrenfeld wryly warns, "ethicists should not do it for us; this process is too important to leave to the professionals."[1] These issues should be discussed by any, and all, persons who are interested in the implications of scientific research and technological development. The importance of dialogue on a local, pre-public policy level is especially pronounced with regard to new technologies. Hudson, Baruch and Javitt indicate, in their essay on prenatal genetic diagnosis, that policy and regulatory structures are not necessarily in place to respond to emergent technologies.[2] Local dialogue about emergent technologies is essential because the implications of these issues are only beginning to enter conversation at national or political levels.

Another premise of this volume is that society's perceptions and concerns are the roots of public policy. As citizens of representative democracies, our opinions can—and should—influence the formation of public policy. To hope that someone (or some government) will solve these very real problems for us is to severely curtail the richness of our political system at a time when the concerns of science and society are increasingly intermingled. Human values, in their unremitting diversity,

[1] David Ehrenfeld, "Unethical Contexts for Ethical Questions," 34 (in this volume).

[2] Kathy Hudson, Susannah Baruch, and Gail Javitt, "Genetic Testing of Human Embryos: Ethical Challenges and Policy Choices" (in this volume).

shape moral debate; and moral debate should help to shape public policy. Since much scientific research occurs without ethically informed or effective regulations, democratic citizens can hardly afford *not* to pay attention to these questions. Gorovitz, Angell, Macklin and Woodwell each observe that the decisions made—or avoided—by individual researchers (or governments) have truly global ramifications. Hence debates over topics such as human subjects research, the availability of reproductive health care, and environmental protection and resource use will continue to be of central importance.

Expanding Horizons in Bioethics thus points the reader toward public policy, but it considers the foundations of public policy to be of enormous consequence. We are concerned with people's values and opinions that are not necessarily inscribed into legislation or established by legal precedent. This aspect of "policy" deserves attention, although it is often implicit, contingent, and far from systematic. It resides, instead, at the earlier level of interest about particular issues, where opinion and action take shape. In this volume we strive to encourage these *roots* of policy, rather than to advocate for a particular policy-based approach to a set of bioethical issues (although some authors in this volume certainly endorse specific public policy actions). Furthermore, essayists from each section of this volume indicate that, in the absence of official public policy and effective regulation, the *de facto* "policy" will simply be the sum of individual human actions and decisions.[3] Since the policies of the future depend on the ethical inclinations and purposeful actions of the present, there is an enormous need to consider carefully the actions that we take now.

Richard Rorty has aptly encapsulated this pragmatic relationship between concrete human practices and generalized public policies. Though he is speaking of pragmatism as a philosophical worldview, his comments are apposite for a society concerned with bioethics:

> The purpose of inquiry is to achieve agreement among human beings about what to do, to bring about consensus on the ends to be achieved and the means to be used to achieve those ends. Inquiry that does not achieve coordination of behavior is not inquiry but simply wordplay. To argue for a certain theory about the microstructure of material bodies, or about the proper balance of powers between branches of government, is to argue about what we should do: how we should use the tools at our disposal in order to make technological, or political, progress ... there is no sharp break between natural science and social science, nor between social science and politics, nor between politics, philosophy and literature. All areas of culture are parts of the same endeavor to make life better. There is no deep split between theory and practice, because on a pragmatist view all so-called 'theory' which is not wordplay is already always practice.[4]

Through these approaches just outlined, we hope to prompt readers to ask questions about the generation of individual (or community) opinion on certain

[3] Gorovitz claims that "[f]or now it seems likely that the 'standards' of use of biotechnology will result more from the cumulative impact of individual decisions than from a collective oversight body" (16). Hudson, Baruch and Javitt assert that "in the absence of agreement [on oversight mechanisms], decisions about when and whether to use [prenatal genetic diagnosis] will continue to be made largely by providers and patients" (121). Finally, Woodwell maintains that "the global environmental squeeze is the global integration of specific local failures around the world" (239).

[4] Rorty, xxv.

issues and, subsequently, the generation of public policy. What are the goods and needs, the strengths and the vulnerabilities, of patients, of society, or of nature? How do we evaluate the potential of scientific discovery, or new modes of therapeutic treatment, or competing concepts of environmental value? And once we have established some form of consensus about such claims, how might we navigate the nuanced terrain of science and society, modern medicine, and the environment? These are the "expanding horizons in bioethics" to which our volume refers.

II.

As the editors of the first volume note, "bioethics" as we know it is generally understood to be the diversified heir of medical ethics. Approximately a generation ago, medical ethics emerged with force. Seminal events included: the Declaration of Helsinki (1964), Henry Beecher's incisive review of twenty-two human experimentation studies in the *New England Journal of Medicine* (1966), the founding of The Hastings Center and its influential periodical (1969), the publication of *The Patient as Person* by theologian Paul Ramsey (1970), the Supreme Court decision to legalize abortion in Roe v. Wade (1973), the Karen Ann Quinlan case (1976), and the publication of the Belmont Report (1978). During this period basic precepts and practices of medicine were reevaluated. Physicians, philosophers, lawyers, theologians, policymakers and others found themselves on the cusp of an interdisciplinary dialogue that forever changed the countenance of Western medicine. New technologies, an increasing diversity and complexity of ethical quandaries, and new challenges to certain priorities and practices in medicine perpetuated these cycles of inquiry.

However, physician conduct, patient autonomy and human subjects research were not the only items on the radar screen during this era. In non-clinical research, for example, the development of recombinant DNA technology raised questions about what types of scientific research could be considered morally permissible. In their Preface to the 1st Edition of *On Moral Medicine,* Stephen Lammers and Allen Verhey recall a prescient observation made in 1964 by Kenneth Boulding. Writing after the discovery of the structure of DNA but before the accelerated pace of discovery that bespeaks our current situation, Boulding predicted that a "biological revolution" would characterize the twentieth century. Lammers and Verhey suggest that this "revolution" has had

> consequences as dramatic and profound as those of the industrial revolution of the eighteenth century … with the help of biological and behavioral sciences, human beings are seizing control over human nature and human destiny. That is what makes the biological revolution 'revolutionary'; the nature now under human dominion is *human* nature. We are the stakes as well as the players.[5]

In 1952 the discovery of the structure of DNA was announced by Watson and Crick in a famous issue of *Nature.* Two decades later, it was established that recombinant DNA could be used to generate quantities of DNA and derivative

[5] Lammers and Verhey, eds., xv. See Boulding, *The Meaning of the Twentieth Century,* 7.

proteins, which could then be used for further research or production of therapeutic substances such as insulin.[6] Thirty years after that, in the early twenty-first century, the human genome had been mapped. In short, it seems evident that genetics is one realm in which a "biological revolution" has transpired. Examples abound of the applications of these genetic advances. We are now able to test human embryos for a host of genetic traits and conditions.[7] Genetically modified seed and food sources have come under the purview of agricultural interests, and in some cases genetic material of one species of organism is inserted into an organism of a different species (often, though not always, these gene exchanges occur within the same genus). Gene transfers can be done in the service of various ends—for example, to conduct research on the possible generation of human organs for transplant, to increase the growth rate of a certain type of fishery-bred salmon, or to enhance the heartiness or succulence of certain fruits.

These scientific advances are revolutionary in another way: they raise questions previously only imagined by philosophers and science fiction writers. For example, genetic testing of human embryos often leads prospective parents to decisions of whether or not they want a future child with certain genetic traits.[8] These decisions in themselves might seem relatively straightforward in the case of heritable genetic diseases, such as Tay Sachs or Huntington's; but soon the diversity of tests may tempt us to identify non-disease-linked traits.[9] These capabilities indeed make it possible to choose the traits of our children, but to what degree is it morally permissible to use these capabilities? Moreover, while genetic research currently enables us to select embryos based on the presence or absence of certain traits, genetic research in the future will likely center on germline modifications that actively alter the genetic constitution of human embryos. If that technology is realized in human reproductive contexts, it could potentially grant us the power to eliminate (or augment) traits of developing embryos; and thus, in a very literal fashion, it would shape what we consider "acceptable" genotypes and phenotypes.

In the nonhuman realm, transfer of genes between organisms leads to questions about the wisdom of mixing genetic material and altering the genotypes of organisms; indeed, these questions have been prominent since the 1975 Asilomar Conference on recombinant DNA technology. Genetic enhancement of certain food

[6] Recombinant DNA technology was a watershed in molecular biology because it enabled scientists to generate materials that cells normally produce, but on a drastically reduced timeframe and with much greater yield. In this procedure, a gene (or series of genes) is inserted into bacteria. Each time that a bacterium divides—which is relatively often—it reproduces the gene. These "mass produced" copies of DNA (or its derived proteins) can then be harvested and used for therapeutic purposes or experimental study. The prospects of the technology and the need for oversight were affirmed at the famous Asilomar Conference of 1975.

[7] See "Genetic Testing of Human Embryos: Ethical Challenges and Policy Choices," by Kathy Hudson, Susannah Baruch, and Gail Javitt (in this volume).

[8] Supra n. 7; see also the essay by Karen Lebacqz, "Choosing our Children" (in this volume).

[9] Some prospective parents already select embryos according to sex. In addition, it should be noted that prenatal genetic diagnosis currently only occurs in conjunction with in vitro fertilization—although Hudson, Baruch and Javitt indicate that this may be subject to change as more genetic tests become available.

sources has been, and continues to be, a contentious international issue.[10] Again we may affirm that the "biological revolution" of which Boulding spoke is particularly evident in genetics. It has changed the scope of our knowledge about the natural world as well as our ability to manipulate and influence it.

Given the plethora of issues entailed in medicine and scientific research, the robust history of debate on these topics, and the obvious fact that humans tend to concern themselves with issues related to human well-being, it is no surprise that medical ethics is a prominent realm of bioethics. As such, medical ethics forms one of our three categories of inquiry in this volume. Nonetheless it is only *one* realm of bioethics, for environmental issues have never been far from mind. In 1949 Aldo Leopold's *A Sand County Almanac* ushered in a new generation of environmental thinking: Leopold is widely credited with introducing the concept that nature in itself has intrinsic value. Soon after, philosophers and ethicists began to persistently reassess notions of the relationship between humans and the environment.[11] Rachel Carson's book *Silent Spring* (1962) famously alerted society to the undesirable effects of widely used agricultural petrochemicals such as DDT. Her questions and indictments popularized debates over the proper use of scientific technologies and biological or chemical compounds when humans and nature are at risk. Ecologists and other scientists contributed to this discussion by envisioning concepts—such as "the biosphere"—that encompassed, and gave new weight to, the reality of the earth's interdependent natural processes.[12]

Most people would agree that ecology, environmental ethics and environmental philosophy are appropriate realms of inquiry in modern bioethics. In fact, environmental concerns are constitutively linked to the origin of the word "bioethics." In *A Companion to Bioethics,* Helga Kuhse and Peter Singer note that the term "bioethics" was initially coined in 1970 by Van Rensselaer Potter to signify "a 'science of survival' in the ecological sense—that is, an interdisciplinary study aimed at ensuring the preservation of the biosphere."[13] Linguistically, at least, it seems that environmental ethics has a solid claim to the bioethical stage.

The most immediate and obvious difference between medical ethics and environmental ethics is the fact that the object of analysis in the latter is not always the human person. Of course, human action or nonaction is always implicated in bioethics. But it does not follow that humans should be the exclusive loci of inquiry. Awareness of human hopes and vulnerabilities is important; so too is attention to groups and communities, to endangered species of organisms, to clusters of natural entities, and to ecosystems. The interaction of science and society, the robust field of medical ethics, and the provocative claims of environmental ethics all deserve a place in bioethical dialogue. In this book we have attempted to give voice to each.

[10] See the discussion of Bt-toxin and genetically modified crops in Ehrenfeld, "Unethical Contexts for Ethical Questions" (in this volume).

[11] See "The Pragmatic Power and Promise of Theoretical Environmental Ethics," by J. Baird Callicott (in this volume). Callicott claims that theoretical environmental philosophy has helped to advance the cause of environmental awareness and policy.

[12] See, e.g., "Science, Conservation, and Global Security" by George Woodwell (in this volume).

[13] See Potter, 127-153. Cited in Kuhse and Singer.

Emergent technologies can change the way that humans live, and—to quote Gorovitz—they "may transform the sorts of creatures that we are."[14] This is true regardless of whether the lens is on environmental ethics or medical ethics; on the particular values of one culture; or on humans as one set of technologically-savvy inhabitants (some would say stewards[15]) of a single biosphere. The fact that such technologies can impact the way humans live within the larger world is why we have chosen the heading "Science and Society" to encapsulate our first set of inquiries in this book. The essays by Gorovitz and Ehrenfeld—the first a philosopher, the second a biologist who also trained as a medical doctor—can be seen as social philosophies that critically engage science. These set the stage for the next three essays in that section. We then turn to medical ethics, and finally to environmental ethics. We think that you will be pleasantly surprised, as we often were, by the consistency of questions woven throughout these three sections of the volume.

III.

Science and Society

The essays in this section introduce questions about the fundamental relationship between science and society, and they comment on the ways in which humans perceive and regulate this interplay. The section begins with broad philosophical claims (Gorovitz) and moves into a discussion of the appropriate contexts for evaluating new technologies (Ehrenfeld). The remaining essays pose the question of research ethics in three distinct ways.

Samuel Gorovitz suggests that human nature will be influenced by the ways that we utilize scientific advancements. He discusses enduring philosophical questions and raises questions about the appropriate use of biotechnological interventions. Drawing from philosophy, literature, social criticism, and contemporary news sources, Gorovitz explores the types of issues that face humans—and concludes that "the need for discernment, attentive listening, reasoned deliberation, and socially responsible advocacy" are crucial if we are to use technology wisely and well.

David Ehrenfeld endorses a "widening of context" as individuals and societies assess various technologies. He discusses the multiple and far-reaching—but often overlooked or ignored—effects of certain technologies, such as the use of recombinant bovine growth hormone in milk. Ehrenfeld claims that we must survey as many contexts as possible in order to fruitfully assess human uses of technology. Moreover, Ehrenfeld emphatically maintains that it is not only "legitimate" to widen the context of inquiry—"[i]t is practically and ethically essential."[16]

[14] Gorovitz, "The Past, Present, and Future of Human Nature," 3 (in this volume).

[15] See, e.g., "The Expanding Circle and Moral Community—Naturally Speaking," by Chalmers Clark (in this volume).

[16] Ehrenfeld, "Unethical Contexts for Ethical Questions," 28 (in this volume).

C. Kristina Gunsalus proposes that human subjects research standards do not translate well into the humanistic disciplines. She suggests that the current concern over clinical human subjects protections has actually impeded research in the humanistic disciplines. Gunsalus traces the history of human subjects oversight from its origins in clinical research ethics, and she explores the ways in which terms such as "research" and "generalizable knowledge"—and even "risk" or "harm"—can have dramatically different meanings across disciplinary boundaries. Her essay captures the tension between protection of subjects and intellectual freedom. She emphasizes the need to reassess regulatory mechanisms (such as Institutional Review Boards) in the humanistic disciplines. If we fail to do this, Gunsalus warns, it will "call our entire regulatory status into disrepute."[17]

Jonathan Moreno describes the U.S. government's legacy of human subjects protections from 1916 to the current era. He claims that the bioethics community has much to learn from this history, particularly since most of this information was only recently declassified. Reflecting on his experience as a member of the federal government's Advisory Committee on Human Radiation Experiments, Moreno traces the formulation of human experimentation standards in military contexts during the twentieth century. He maintains that the government's policies on human subjects research provide an interesting lens into human subjects protections, and that the lessons learned may be timely and relevant for current issues facing our society.

Marcia Angell focuses on the question of researchers' responsibilities to human subjects, particularly when the research is conducted offshore and cross-culturally. She retrospectively investigates the well-publicized AZT (zidovudine) trials in Africa during the 1990s and delineates opposing sets of arguments about researchers' obligations to persons enrolled in their studies. Angell—who was the executive editor of the *New England Journal of Medicine* during this time period, and who found herself embroiled in the ethical controversies—stipulates several moral mandates and procedural recommendations about research on human populations. She raises troubling questions about the current regulatory structure, discusses some of the "incentives" toward offshore research, and suggests that individual researchers and policymakers have a large responsibility to minimize (and, if possible, to eradicate) unethical conduct in research involving human subjects.

Medical Ethics

Ruth Macklin claims that women's access to effective reproductive health care in the developing world is severely curtailed by cultural resistance to reproductive health practices, variation in medical treatment, and limited availability of resources. Macklin articulates how competing value systems—cultural, religious, and political –impinge upon women's reproductive rights on local, national and international

[17] Gunsalus, "Human Subject Protections: Some Thoughts on Costs and Benefits in the Humanistic Disciplines," 57 (in this volume).

levels. Macklin contends that reproductive rights are a fundamental part of human rights, and that a denial of the former is an abnegation of the latter.

Kathy Hudson, Susannah Baruch and Gail Javitt discuss genetic testing of human embryos. They describe current procedural aspects of prenatal genetic diagnosis (PGD), raise ethical questions related to this technology, and describe the policy frameworks that might be employed to regulate the use of PGD. In addition, the authors identify areas where further research is necessary, and they attend to questions of justice (such as access to and affordability of PGD).

Karen Lebacqz intensifies refrains from the preceding two essays by asking, in the context of reproductive cloning, what the extent of our reproductive rights should be. Her essay is a response to provocative claims made by lawyer John Robertson, who holds that a right to procreate entails the right to choose the characteristics of one's offspring. In her critique of this notion, Lebacqz problematizes several explicit and implicit assumptions in Robertson's logic. Along the way she launches important questions of moral permissibility that are grounded in considerations of justice and feminism.

Harry Rosenberg, a medical statistician, reveals the potentially misleading nature of health statistics with respect to debates over heart disease in the late 1990s. Using census and death certificate data, Rosenberg shows how four types of statistical indicators of mortality can be variously interpreted and portrayed to the public in ways amenable to the motives of government and public interest groups. Rosenberg maintains that while good intentions may undergird the selective use of statistics to convey a particular message, health statisticians nonetheless have a responsibility to make certain—so far as possible—that the information they generate is used responsibly, so that the public's perception of health trends is appropriate.

James Flory and Ezekiel Emanuel describe the ways in which different demographic groups within the United States have varying levels of access to, and experiences of, end-of-life care. Their chapter attempts to "introduce the most pressing policy and ethical issues, in the context of the major trends and innovations in end-of-life care over the last thirty years of the twentieth century."[18] They describe core medical, legal, and ethical developments and note economic concerns and disparities in care. Ultimately, Flory and Emanuel identify issues that will require the attention of policymakers in the near future if we are to care well for those who are dying.

Environmental Ethics

Writing in defense of the pragmatic efficacy of theoretical environmental philosophy, **J. Baird Callicott** claims that concepts envisioned by philosophy—such as "human rights" or "intrinsic value in nature"—impact practice in dramatic ways. He provides an overview of the core movements in the intrinsic-versus-instrumental

[18] Flory and Emanuel, "Recent History of End-of-Life Care and Implications for the Future," 170 (in this volume).

value-in-nature debates over the past several decades. Callicott endorses the pragmatic viability of intrinsic value-in-nature through philosophical argumentation. In addition, he indicates significant ways in which the concept of intrinsic value-in-nature has influenced environmental action.

Chalmers Clark poses a naturalistic argument in defense of intrinsic value-in-nature. Following Peter Singer, Clark argues that it is by no means clear that humans' moral obligations are constrained to the species *Homo sapiens.* He also endorses W.V. Quine's naturalistic view that humans are genetically constituted to care for others. Clark claims that a stewardly relationship between humans and nonhuman natural entities might be a way to affirm this tendency toward altruism. He discusses the role that "trust" might play in such a relationship, which raises problems for the notion of stewardship.

George Woodwell addresses the question of environmental degradation from the perspective of one who is deeply experienced in the intricacies of environmental policy and advocacy. He holds that two complementary courses of action are necessary if we are to retain the integrity of the earth and its inhabitants: first, to reduce our dependence on oil and other fossil fuels; and second, to actively "restore the physical, chemical and biotic integrity of the biosphere."[19] In arguing for these two theses, Woodwell chronicles major impulses, advancements and setbacks in the formulation and implementation of global environmental policy. He offers specific recommendations for ethically sound environmental action. Further, Woodwell proposes that environmental integrity is an essential component of global political stability, particularly as environmental crises become more acute and widespread.

David Goodstein, a physicist by training, concludes the volume with the premise that energy is a necessary condition of modern human existence—and that we are literally "running out of gas." Using M. King Hubbert's prognostications about the limited availability of oil, Goodstein claims that there is an overwhelming need for protracted investigation into alternative sources of energy. He describes several different potential sources of energy, suggesting the benefits and drawbacks to each. Goodstein's conclusion, like so many others in this volume, is disarmingly simple and alarmingly intense: "What happens next is up to us."[20]

All of the essays in this volume recognize that where there is knowledge and technological ability, there is often use of that knowledge and ability. It is fairly well assured that we are influencing the terms of existence of many inhabitants of this planet, from flora to fauna to humans. Moreover, history has shown that while technologies can be used neutrally, they can be (and have been) used to the great benefit, or the great detriment, of human life and the flourishing of the world as a whole. How various types of knowledge and technological ability will be deployed is up to us, individually and collectively. How such information and ability *should* be deployed, and for what reasons, are questions at the core of bioethical inquiry.

Through the fourteen modern quandaries delineated in *Expanding Horizons in Bioethics,* we strive to underscore the importance of individual and collective

[19] Woodwell, "Science, Conservation, and Global Security," 231 (in this volume).

[20] Goodstein, "Energy, Technology and Climate: Running Out of Gas," 264 (in this volume).

education, opinion, and involvement in bioethical issues. Many of the authors, across all three categories of inquiry, claim in no uncertain terms that the future is up to us.[21] Collectively, the essays in this volume clamor for individuals, groups and societies to acknowledge our power—and our responsibility—to choose wisely in our use of promising technologies. We hope that this volume impels, and to some degree enables, individuals to reflect upon pressing issues in bioethics and the methods by which ethical consensus may be reached. This is the practice of policy writ large, the "expanding horizons in bioethics," about which this volume is most concerned.

[21] See the essays by Gorovitz, Ehrenfeld, Gunsalus, Angell, Hudson et al., Flory and Emanuel, Woodwell, and Goodstein.

REFERENCES

Boulding, Kenneth. *The Meaning of the Twentieth Century.* New York: Harper and Row, 1964.

Galston, Arthur W. and Emily G. Shurr (eds.). *New Dimensions in Bioethics: Science, Ethics and the Formulation of Public Policy.* Dordrecht: Kluwer Academic Publishers, 2001.

Kuhse, Helga and Peter Singer. "What is Bioethics?" In *A Companion to Bioethics.* Edited by Helga Kuhse and Peter Singer. Oxford/Malden: Blackwell Publishers, 1998.

Lammers, Stephen and Allen Verhey. "Preface to the First Edition." In *On Moral Medicine,* 2nd Edition. Edited by Stephen Lammers and Allen Verhey. Grand Rapids: Eerdmans, 1998.

Leopold, Aldo. *A Sand County Almanac, and Sketches Here and There.* New York: Oxford University Press, 1949.

Pojman, Louis (ed.). *Environmental Ethics: Readings in Theory and Application*, 3rd Edition. Belmont: Wadsworth/Thomson Learning, 2001.

Potter, Van Rensselaer. "Bioethics, science of survival." *Biology and Medicine* 14 (1970): 127-53.

Ramsey, Paul. *The Patient As Person.* New Haven: Yale University Press, 1970.

Rorty, Richard. *Philosophy and Social Hope.* London: Penguin Books, 1999.

Steinbock, Bonnie, John Arras, and Alex John London (eds.). *Ethical Issues in Modern Medicine*, 6th Edition. Boston: McGraw-Hill, 2003.

I.

SCIENCE AND SOCIETY

SAMUEL GOROVITZ

THE PAST, PRESENT AND FUTURE OF HUMAN NATURE

We live at a time of multiple revolutions in science, particularly in medical science and technology. Profound innovations are emerging that will not just change the problems we address or the ways we respond to problems, but which may transform the sorts of creatures that we are. Our species may be changed for better or worse by what is in prospect. The nature of the human species is therefore at stake.

"*The Past, Present and Future of Human Nature*" might seem all encompassing, but it leaves almost everything out. For most of the time that living creatures have existed there have been no humans and thus no human nature. We are a recent and probably transient phenomenon; we are well advised to keep that humbling fact in mind. Still, as a member of this recent but disarmingly clever species, I have an interest in understanding what our nature is, what it might become, and how that might depend on the choices we make.

It is distinctively human to engage in self-conscious reflection on our own nature. Doing so has gone on for all of recorded human history and must have been going on longer than that. Wonder is among the most salient distinguishing characteristics of humans: we are the self-reflective creature. Our intellectual history centers prominently on efforts to understand our own behavior and motivations, our relationship to nature, and our place in the universe or, on some views, beyond.

We have devoted much thought to exploring whether we are by nature altruistic or selfish, warlike or peace-loving, monogamous or polygamous, shaped more by genetics or by experience and environment, driven by deterministic causes or free to make autonomous choices. We have sought answers to understand the human condition, and to know whether we are the result of purposive design or the chance product of natural processes. We have long wondered about the relationship among our minds, our brains, and the baffling phenomena of consciousness, personal identity, and self-awareness. These issues were historically addressed primarily by philosophers and theologians; now, they are also vigorously pursued by others such as sociobiologists, cognitive neuroscientists, computational linguists, and physicists. We have sought to discern what about ourselves is inherent in our nature and what is socially constructed. We find ourselves unendingly fascinating.

On the traditional Judeo-Christian view, humans were the apex of God's intentional creation; distinct from the rest of the world, they exerted dominion over nature. Other views, such as those of some Native American cultures, saw humans as being at one with nature, properly seeking harmony with the larger whole. Science, even in its earliest iterations, is the human effort to understand what nature

A. W. Galston and C. Z. Peppard (eds.), Expanding Horizons in Bioethics, 3-18.

is and how it works. The history of science is partly a history of changing conceptions of humans in relation to the rest of nature. That history is also one of sequential losses of innocence.

I.

In his *Essay on Man*, Alexander Pope counseled, "Know then thyself, presume not God to scan. The proper study of mankind is man."[1] And he called our attention to the great chain of being – a chain in which he saw us as a middling link. Now even *that* solace is gone; our understanding of evolution has disrupted our notion of progress, for it shows us that we are subject to changes that we may or may not like.

Similarly, those who embraced a geocentric worldview ascribed to the self-indulgent misperception that human life on earth was the point and purpose of the universe's processes. But the geocentrism that Galileo was punished for challenging soon gave way to the scientific evidence that proved him right. However, even after conceding that we were not literally the center of the universe, we wondered about our origin and saw human life as a singularity within the living world. Darwin's genius was his ability to illuminate those mechanisms of natural selection that led to our evolution, and which linked us irrevocably with the processes of the rest of the living world. For a while thereafter, we were content to see ourselves as the product—and to some degree, the goal—of directed evolution. But we no longer have reason to believe that the universe is benevolent or caring, that our emergence as a species is the "intention" of evolution, or that our place in the world is more than the result of the combined effects of the things that happen to us and the things that we do.

I do not claim that the universe is hostile, as suggested by Melville's remark that:

> There are certain queer times and occasions in this strange mixed affair we call life when a man takes this whole universe for a vast practical joke, though the wit thereof he but dimly discerns, and more than suspects that the joke is at nobody's expense but his own.[2]

I claim only that there is no adequate reason to think of the universe as anything but indifferent to human affairs.

II.

So much for the past. What of the present? The old myths and explanations are largely gone, but the old questions and puzzlements linger, compounded by new anxieties as we contemplate an intimidating future. Detached from past moorings,

[1] Pope, Epistle ii, Line 1.
[2] Melville, 214.

we are uncertain about the future of our earth and our species. We are unsure how to understand our existence.

The French existentialist Jean Paul Sartre wrote that humans are nothing but what they make of themselves. He overstates the case: surely part of our human nature is determined by what we are as material objects—chemically complex, biologically living beings with properties we have not chosen and which, until recently, we have not been able to alter. Yet, Sartre's point about human nature has real force.

Sartre claims that for humans, existence precedes essence, and he contrasts human beings as volitional agents with human artifacts. Artifacts come into being when humans, facing a new task, envision a tool to facilitate the performance of that task—such as a hook or knife or mallet to suit our purpose. Then, having first had this idea—an essence—in mind, we create the object to correspond with that idea or essence. Because we first envision the nature of the object, its essence comes first logically. We then create it, so its existence follows its essence.

Sartre says that humans are not like that object. We exist and perform actions. By acting, we make it true that we behave in certain ways. This creates our nature, our essence, which follows from our existence and our actions.[3] There thus can be no inhuman act performed by any human, however grotesquely that behavior may violate deeply held values—be it by Nazi murderers, serial killers, suicide bombers, child abusers or the like. For Sartre, our actions do to some extent define human nature, because existence precedes and *dictates* essence. On this view human nature is limited primarily by the constraints of physical realities. And not only do we create our individual natures but also that of our species, as the sum of individual actions. What we choose to do individually shapes human nature collectively. We cannot avoid this responsibility, for to live at all is to act and thus to contribute to the evolution of human nature.

That we partially determine our own nature is not news. But in this era of such developments as pre-implantation genetic screening, computerized prosthetics, and stem cell research, we see exhilarating and disturbing new powers, possibilities, risks, and responsibilities. As we influence the course of human nature through our use of new technologies, what does the future hold for our species? The future, in the sense relevant here, is not next week or next year. Some consequences may become evident over several decades; most will become evident a few hundred years later, or more. Our influence will be felt millennia from now, perhaps even more powerfully than we are now influenced by the enduring heritages of Hippocrates, or Aristotle, or the authors of the Bible. We therefore must ask what both the short and long term consequences of our actions may be.

[3] Jean-Paul Sartre writes: "If man, as the existentialist conceives him, is indefinable, it is because at first he is nothing. Only afterward will he be something, and he himself will have made what he will be. Thus there is no human nature." *Existentialism and the Human Emotions*, 15.

III.

As has often been noted, the natural pace of biological evolution is slow in comparison with the much faster pace of cultural change. There is a radical difference between the pace of scientific discovery and the pace of the decisional deliberation that is needed to guide the wise uses of scientific discovery. We need many forums of inquiry, and strong advocacy for the importance of unhurried and broad-based reflection. We run the risk of both over regulation and under regulation, of inventing capacities we lack the wisdom to handle responsibly, or being so wary of what research can yield that we foreclose the prospects of those very developments that can empower us to thrive. In the end, what research is done and used will reflect public sentiment. Whether those public judgments are based on accurate understanding, confusion, or irrationality is up to us. Public education about science and technology, and the processes of public choice, must therefore be among our top priorities as potentially life-altering technologies emerge.

In 1962, at a conference in London on Man and His Future, Sir Julian Huxley spoke on *The Future of Man – Evolutionary Aspects*, affirming that evolutionary change is inevitable, that the human species will continue to evolve. And he said:

> If blind, opportunistic, and automatic natural selection could conjure man out of a viroid in a couple of thousand million years, what could not man's conscious and purposeful efforts achieve even in a couple of million years, let alone in the thousands of millions to which he can reasonably look forward?[4]

Similarly, the great geneticist Theodosius Dobzhansky wrote:

> Man, if he so chooses, may introduce his purposes into his evolution… He may choose to direct his evolution toward the purposes he regards as good… The crux of the matter is evidently what purposes, aims, or goals we should choose to strive for. Let us not delude ourselves with easy answers.[5]

However, Dobzhansky did not provide a map revealing that direction.

We should remain humble both about our technical and scientific understanding, and about our capacity to understand what will, in the long term, best serve our interests—what direction will be "good and desirable." In particular, we should avoid the perils of perfectionism in contemplating our future. There is typically a tension between the pursuit of improvement (one main purpose of scientific inquiry) and the risks of being over-zealous in that pursuit. Part of the problem is our inability to know in advance what we will later count as an improvement, rather than just a change or even a decline. Another part is our inability to anticipate accurately all the perils of zealous pursuit.

One illustrative example is the quest for efficiency. Efficiency, good for many reasons, has become the profit-maximizing deity for American commercial corporations (and far too many universities and hospitals). Striving for perfect

[4] The CIBA Foundation sponsored the London conference entitled "Man and His Future" on November 27-30, 1962. Sir Julian Huxley opened the conference with his lecture, "The Future of Man: Evolutionary Aspects," which was later published in Wolstenholme, ed., *Man and His Future.* Huxley's speech appears on pages 20-22.

[5] Dobzhansky, 163.

efficiency, however, has brought some enterprises to the brink of collapse—or beyond—when they lack the flexibility to respond to unexpected developments. They have become too specifically adapted to their niche to survive environmental change. What appears to be lack of waste or slack in an organization can abruptly emerge as an inability to respond to unanticipated constraints or opportunities.

A second example is quality control. Flaws in production and errors in judgment are bad. But the quest for error-free functioning, or for zero defect manufacturing, can have costs beyond bearing and barriers beyond surmounting. A zero error goal is unreachable in any substantial and complex domain, including those in which the stakes are vitally high—such as cardiac surgery or, alas, space flight. And it will be unreachable as we try to eliminate disease, extend life, have ideal children, and enhance human capacities. Improvement is typically a worthy objective, but the demand for perfection is dangerous because it does not account for the complexities of real life and human action.

People are already engaged in the effort to determine their descendants' characteristics. As we consider how to try to shape the future of our descendants and our species, we should remember that the quest for perfection has perils often under-appreciated in advance. Leon Kass put the point well in reviewing a book entitled *Brave New Worlds*, a play on the title of Aldous Huxley's novel *Brave New World*. Citing Aldous Huxley, Julian Huxley's brother, Kass wrote:

> Huxley knew that it is generally harder to recognize and combat those evils that are inextricably linked to successful attainment of partial goods. The much pursued elimination of disease, aggression, pain, anxiety, suffering, hatred, guilt, envy and grief, Huxley's novel makes clear, comes unavoidably at the price of homogenization, mediocrity, pacification, drug induced contentment, trivialized human attachments, debasement of taste and souls without loves or longings – the inevitable result of making the essence of human nature the final object of the "conquest of nature for the relief of man's estate," in the words of Francis Bacon.[6]

Sartre is instructive here as well. He emphasized that we are free to choose what we do; he also stressed that whenever we are free to make a decision, we are unavoidably required to make it. We may delude ourselves in an effort to avoid the anxiety engendered by acknowledging our freedom, but there is no escape from the responsibility of making a choice. We do not have the option of watching our own evolution play out as if it were not in part our own doing.

Much is at stake that we do not now and never will fully understand, and that we cannot fully control. The research that helps us learn so much about what we are as living organisms also can empower us to change what we are, *not just as physical beings but also as human moral agents*.

[6] Kass, "Beyond Biology." It is ironic that Kass, whose cautionary insights suggest a subtle and rich wisdom, is now widely seen as a political operative driven more by ideology than by open-minded inquiry. This evaluation of Kass is catalyzed by his manipulative (and, some have argued, duplicitous) management of an assessment of stem cell research in *Human Cloning and Human Dignity: The Report of the President's Council on Bioethics*, which supplants the earlier recommendations of The National Bioethics Advisory Commission (which President Bush elected to eliminate before creating the new President's Council on Bioethics, chaired by Kass).

IV.

It is not clear how best to proceed. Assessing projected risks and benefits of various technological advances will be dauntingly difficult given the range of opinions and needs in society. Oversight of these technologies is essential, however, if we are to have a reasoned influence on the consequences of their development. Some of the decisions will play out in the private choices individuals make about how to use new reproductive options. Other individuals will benefit from new therapeutic choices for chronic or acute illness, such as occurred in November 1998, when the first successful therapeutic use of genetic therapy was reported in patients who began to grow new blood vessels after their hearts were injected with a gene that makes a protein called vascular endothelial growth factor. For now it seems likely that "standards" of use of biotechnology will result more from the cumulative impact of individual decisions than from a collective oversight body.

Yet we also face collective choices about when and how to manage the uses of our new capacities, and which technologies should be sanctioned in the first place. Who should be responsible for such oversight, and how it should be administered, are unresolved questions. Should the federal government regulate emerging technologies; or should individuals be able to use technologies as they wish, so long as the effects do not impact others too strongly? As the enduring debate about physician-assisted suicide demonstrates, we are not even clear about how to distinguish between a private domain in which individual autonomy is a dominant value, and the domains in which public constraint is appropriate. There are no clear maps to an ideal moral future.

Resolving these matters is not a challenge for science or for technology alone. It is a question of fundamental values. In saying that questions of fundamental value are not questions for science or technology, I affirm only that they cannot be resolved by science or technology alone. Something more is needed. Of course, ethical reflection must be carefully and honestly grounded in the empirical facts that give rise to ethical puzzlement in the first place. Ethical judgments that are scientifically naive lack credibility. But science is not enough, and we can understand why. Science is driven in part by practical goals, but is not only—perhaps is not even primarily—about learning to control various aspects of nature for practical purposes. It is also about the intellectual and aesthetic satisfactions that enlarge the human spirit when we confront the mysteries of nature and achieve some success in comprehending them. What we have learned about DNA and the evolution of biological organisms has revealed much about the mechanisms of life. We know, roughly, an outline of basic genetic functioning. We do not know much about how this complex system got started (though conjecture of various degrees of plausibility abounds), and there is immense detail that we do not understand about how it works.

We learn much from science as it responds to puzzlement and mystery by investigating, seeking evidence, hypothesizing, and interpreting—but it has limits, some of which we readily acknowledge. For example, we cannot know the interior of black holes—what physicist David Goodstein has called "the naked singularity inside where the known laws of physics break down."[7] We cannot know the boundaries of deep space, or accurately predict heart attacks or the trajectories of asteroids, or even know with certainty if particular genes will be expressed in a human.

The development of powerful new capacities and understandings in a full range of domains, including information processing, neurological research, and genetics, may distract us from the realization that all of this together constitutes only a minuscule amount of what we would have to know to control the direction of human evolution in reliably predictable ways. Despite genuinely impressive developments in human understanding of human beings, we ought not lose awareness of our own intellectual fallibility. We do not know, and we are not likely ever to know, how to resolve the larger mysteries of our human experience of life, full of hopes and fears, aspirations and disappointments, joys and sorrows, pride and embarrassment, loneliness and solidarity, generosity and selfishness, and all the rest that constitutes the mysterious core of our inner lives.

V.

Our challenge is to find clarity and ultimately agreement about the criteria by which we judge whether evolutionary changes in humans are positive. If science alone is not enough to evaluate the import and value of such changes, what then of religion, which also offers views of human nature and of what paths it is best to pursue? I do not here refer to religion as a complex and pervasively influential social practice—sometimes vitally helpful, sometimes lethally destructive, and usually somewhere between. I refer to the theological—as opposed to social or cultural—part of religion that affirms propositions about the nature of humanity in relation to a larger reality. In some of its forms, such aspects of religion offer to dispel enduring uncertainty about mysteries that matter deeply to us, but which refuse to yield to scientific inquiry. Religion, in this respect, sometimes comforts those who feel that a need for answers is more powerful than their need for scientific evidence (although not all religious perspectives are opposed to unresolved uncertainty with regard to the natural world). On any controversial ethical issue, the full range of positions can find support within one religious tradition or another—and often within the various perspectives that flourish in a single tradition. Religious perspectives can enrich our moral sensitivities, but without science they can neither resolve our ethical quandaries nor fill the gaps in our understanding of the universe, no matter how much we yearn for such clarity. Rationality may require that we suspend judgment even about very important mysteries when we lack credible evidence about them.

[7] Goodstein, personal communication.

VI.

Given that neither science nor religion suffices to guide our evolutionary future, perhaps we should consider the mechanisms of public choice. These include more than just government. Social opprobrium, for example, has played a major role in the remarkable reduction of drunken driving over the past few decades and in the transformation of public attitudes toward smoking. But social opprobrium is not the whole story; it influences and is strengthened by the mechanisms of government, which is, after all, the primary instrumentality of public choice. It can oppose a mode of behavior by vigorously or mildly enforced legal prohibition, by the provision of disincentives such as taxation or onerous regulations, or by programs of advocacy and public education. Government policies can ignore, investigate, or resolve an issue. So when we consider public choice about emerging genetic powers, we must ask both whether the government should play a role, and what sort of role that ought to be. Indeed, one can be imaginative in interpreting a role.

Consider the Food and Drug Administration's actions during the late 1990s in the matter of smoking as the nation's most serious public health problem. Nothing in the FDA's mandate clearly gave them responsibility for regulating the content, labeling, advertising, or sales of tobacco. Nonetheless, the FDA, with Commissioner David Kessler at the helm (from 1990–1997), knew that the tobacco issue was not being addressed effectively elsewhere; they knew that the tobacco industry was systematically addicting the nation's youth without regard to the disease and death that would follow; they suspected (and later had evidence to support their hypothesis) that cigarette companies knew the addictive nature of nicotine and were systematically increasing the amounts of nicotine in cigarettes; and, finally, the FDA hoped that the bold action of putting smoking on the national health agenda might make a positive difference in health outcomes. So Kessler and the FDA pursued a controversial, risk-taking role expansion in the public interest. Kessler notes that the FDA was trying to prove that nicotine is an addictive drug, that cigarette companies knew about its addictive properties, that cigarettes are drug distribution devices, and hence that cigarettes (and thus cigarette companies) come under FDA regulatory domain.[8]

The position taken by the FDA was utterly defeated in the Supreme Court's five to four ruling in *FDA vs. Brown & Williamson et al.* that the FDA lacks jurisdiction in this domain. Yet what the agency did played an important role in shaping public attitudes, in enhancing the resolve of health officials, in emboldening holders of public office, and in accelerating the moral isolation of a legal but evil and duplicitous industry. The FDA lost in the courts, and though Congress did not address the issue as the FDA commissioner had hoped, the FDA won a major victory nonetheless.

The point is not about tobacco, but that we each have the capacity to contribute to the betterment of the human condition—not just by addressing problems that impinge on our personal lives or which arise in whatever professional roles we fill, but by remembering that our individual humanity empowers and requires us to act

[8] See Kessler, *A Question of Intent,* for an intriguing account of this process.

on behalf of, and in concert with, all humanity. This can be done by actively joining the debates about the issues that will shape humanity's future.

A Model For Discourse

One possibly useful model for those debates is the New York State Task Force on Life and the Law, established in 1985 by Governor Cuomo to recommend public policy on morally controversial aspects of the life sciences. Unlike those bodies created to address specific topics, this is a standing body with no political or ideological slant, and it has the luxury of deliberating in private and at whatever pace it judges is required by a topic. As scientific and technical advances rush toward us at a dazzling pace, there is no substitute for calm, unhurried, reasoned discussion about how best to deal with its consequences. That takes considerable time, and it takes even more time to see the results of such discussion translated into action.

In 1994, in a volume entitled *When Death Is Sought*, the Task Force called for revision of New York State's controlled substance laws in order to make better pain management available to grievously ill patients.[9] That small and relatively uncontroversial recommendation was finally signed into law four years later, in August 1998. In April 1997, the Task Force had released a massive report on *Assisted Reproductive Technologies*, the product of two years' effort by the members and professional staff.[10] This report was described the next day, in a front page *New York Times* article, as "a landmark report" that "sets the standard for further discussion of the issues in this country."[11] Despite the years of careful work and debate by the Task Force, however, and despite the glowing reception of its report, few of the questions at issue have been definitively resolved. That "further discussion" continues vigorously.[12] And so it will and must remain, as the full range of ethically challenging issues in biotechnology resists rapid or easy resolution, legislation or regulatory remedy.

Consider these selected issues stemming from advances in biotechnology.

Reproduction

A *New York Times* article in 1998 bore the headline "With Help of Science, Infertile Couples Can Even Pick Traits."[13] Fertile couples, too, have an interest in influencing, and perhaps in choosing, some of the traits of their children in order to eliminate what they see as defects, and perhaps to provide enhancements that they see as desirable.[14] They too will turn to scientific research to increase their capacity

[9] New York State Task Force on Life and the Law, *When Death is Sought.*

[10] New York State Task Force on Life and the Law, *Assisted Reproductive Technologies.*

[11] Altman.

[12] See, e.g., two essays in this volume, "Genetic Testing of Human Embryos" (by Hudson, Baruch and Javitt) and "Choosing our Children" (by Karen Lebacqz). *Ed.*

[13] "With Help of Science, Infertile Couples Can Even Pick Traits," *New York Times.*

[14] See note 12. *Ed.*

to have the kind of children they most want—perhaps, perfect ones. This is fraught with risk.

Genetic selection is not a new idea. Plato and Aristotle, among others, conjectured about how society might influence reproductive behavior so as to serve the betterment of society—at least as they judged what was better. Plato envisioned a Board of Matrons to supervise young couples because "the best men must have sex with the best women as frequently as possible, while the opposite is true of most inferior men and women."[15] Aristotle favored eugenic methods aimed at promoting socially desirable mating. But they had few effective tools at their disposal for influencing human reproduction and its consequences. We, on the other hand, anticipate a future full of ways to influence the particularities of human reproduction, from the advent of the birth control pill to the uncharted landscape of preimplantation genetic diagnosis to future germline modifications.

Louise Brown, the first person conceived by the use of in vitro fertilization, is in her mid-twenties. In 1978, she was an anomaly; by now, assisted reproduction has led to many tens of thousands of births in many countries. We not only create embryos *in vitro* for later implantation, we screen genetic material to select for desirable properties before fertilization, and we can even anticipate substantial ability to modify that material along the way. Such capacities for influencing our reproductive futures have increased with remarkable speed and may have as yet unknown effects on the child born as a result.

Genetics

Further, genetic engineering offers the possibility—not next week, but in our lifetimes—of manipulation at the molecular level of genetic, heritable phenomena, including the modification of germ line cells. Efforts have been made to patent human genetic materials created in genetics laboratories; some countries have legally prohibited such patents. This issue indicates the intricacy with which complex matters of law, business, and biology have now become inextricably intertwined. This is cause for such concern that the United Nations Educational, Scientific and Cultural Organization adopted a Declaration on Human Dignity and the Human Genome, article one of which states:

> The human genome underlies the fundamental unity of all members of the human family, as well as the recognition of their inherent dignity and diversity. In a symbolic sense, it is the heritage of humanity.[16]

That heritage is more fragile than we have realized until recently. In the early days of evolutionary theory, some commentators, such as Lamarck, thought that acquired traits could be inherited by future generations. But it soon became widely agreed that this was not so, and Lamarck was dismissed as silly for having thought otherwise. That impossibility long remained a basic tenet of evolutionary theory.

[15] Plato, 134.

[16] United Nations Educational, Scientific and Cultural Organization (UNESCO), General Conference, 29th Session.

Now, even this cherished precept of genetics will be lost if genetic therapy moves beyond somatic intervention to germline interventions that affect the genetic character of future generations. In this sense, germline alterations are acquired traits that *will be heritable.* Lamarck may be vindicated after all.

Genetic engineering is only one example of a technology that challenges our sense of what we are and what we might become. Indeed, those challenges are so great and numerous that it makes sense to ask not only what the future of human nature is, but whether there will continue to be a human nature at all. That is the risk I emphasize here.

There is a wondrous inventory of genetic material to tempt us, evidenced by the rich array of living creatures that form the interconnected and interdependent fabric of life. As we have discovered more about the genetics of various organisms, we have learned with some surprise about the underlying commonality of all life, and thus about the full range of our relatives. Our genetic composition overlaps with that of the chimpanzee by more than 99%; they are very nearly us. Even plants share a significant part—on the order of 40%—of our genetic code, and we are not as far removed from yeast as we would have guessed. Consider the possibilities for unraveling all this genetic architecture and recombining it in imaginative ways, and you get a dizzying sense of what could await us as a result of what we have learned through sophisticated research.

In 1952 Crick and Watson, with notoriously under-appreciated participation by Rosalind Franklin, discovered the molecular structure of DNA.[17] In 2003 the National Institutes of Health's (NIH) Human Genome Project announced that they had finished mapping the entire human genome, and in private and federally funded laboratories the mapping continues for genomes of other species. This took nearly fifty years; but the current rate of discovery and scientific change is rapidly accelerating, as evidenced by the controversy over stem cell research.

Stem Cells

In August 1998, *Technology Review* published an article about an obscure research program aimed at identifying the stem cells in human embryos—that is, cells prior to differentiation. The article's heading stated that this "could revolutionize medicine. But to reap the rewards, biotechnology firms and researchers must brave a firestorm of controversy."[18] This research holds the promise—or the threat—of the capacity to control the differential development of human cells so as to produce replacement organs or characteristics of a particularly desired sort. And, as the article pointed out, it also "could conceivably provide a vehicle for the genetic engineering of people." In May 1998, the University of Wisconsin hosted an NIH-sponsored international conference on stem cell research. At that point in time stem cell research was considered experimental, and it was conducted on a broad range of non-human animals. Despite the availability of speakers who were eager to discuss

[17] Cf. Maddox.
[18] Regalado, 34-41.

the implications of human stem cell research, no presentations on human stem cell research were allowed—the subject was too inflammatory. The *Technology Review* article concluded, however, that "[i]t's only a matter of time before it bursts out from behind closed doors and begins to transform the public debate over biomedical ethics and, perhaps, much of medicine as well."[19]

A lead story in November 1998 in newspapers across the country validated that prediction. The science writer Nicholas Wade put it thus:

> Pushing the frontiers of biology closer to the central mystery of life, scientists have for the first time picked out and cultivated the primordial human cells from which an entire individual is created.[20]

The American federal government, despite many positive recommendations by its own deliberative bodies, supports no research involving the use of human embryos. As a result, stem cell research is entirely supported by and located in private industry, which seeks to develop and promote its commercial applications for profit as rapidly as possible. But in January 1999, Harold Varmus, then Director of the NIH, announced that the general counsel of the Department of Health and Human Services had determined that research on human embryonic stem cells does not constitute research on human embryos, that such research therefore would not be subject to the ban on federal funding of human embryo research and thus could be funded by NIH. The National Bioethics Advisory Commission concluded that research involving the derivation and use of embryonic stem cells and embryonic germ cells "has the potential to produce health benefits for individuals who are suffering from serious and often fatal diseases."[21] *Science Magazine* cited such research as the scientific breakthrough of the year for 1999, noting that more than any other development, it meets the criterion of being "a rare discovery that profoundly changes the interpretation or practice of science or its implications for society."[22]

We now have a raging dispute about this line of research, just as predicted in 1998. The federal government maintains that so long as stem cell research proceeds on selected *extant* stem cell lines, federally funded research can proceed; but it bans research on stem cells created specifically for research purposes. The research community and multiple advocates for patients with chronic illness hope for effective treatments derived from stem cell research, even while agreeing that cloning is not an acceptable route to the development of a person (hence the less controversial prohibition of reproductive cloning). Stanford University has a new research initiative, funded without federal support, to pursue stem cell research in the private sector. The press covers this dispute almost daily. The pace of change and the vigor of the conflict in this domain are astonishing. Of course, biotechnological frontiers extend well beyond the realm of cellular biology.

[19] Ibid.
[20] Wade, A-1.
[21] Varmus.
[22] Vogel, 2238-9.

Physical and Mental Enhancement

Consider these few examples: Artificial intelligence, though primitive in respect to the aspirations that many of its early advocates expressed, nonetheless has developed impressive capabilities. What might it portend for the human-machine interface? We have the capacity to provide people who need prosthetic devices with a remarkable array of substitutes for normal human functioning; how might this be extended via biotechnology?

This is no idle speculation. The advocates of self-directed evolution of the human body display their enthusiasms in varied and vigorous ways, many of which can readily be seen at http://www.betterhumans.com. And the "Transvision 2003 USA" Conference at Yale University had as its agenda exploring the World Transhumanist Association's view that "technology can be used to overcome the limitations of the human body, and that individuals should be allowed to enhance their bodies."

In 1988 in France a man who had lost an arm received the world's first hand transplant. At least seven more hand transplants have been reported, in addition to four instances of double hand transplants.[23] As if that were not enough to ponder, in late 2002 British plastic surgeon Peter Butler affirmed that emerging surgical techniques would allow a full-face transplant to be performed imminently; he called for debate about the propriety of doing so.[24] A further question arises: If we can replace a man's lost arm either with a mechanical device or with a transplanted human arm, why not use a better one that does not merely replace what he has lost, but which grants him powers he did not previously have?

The prospect of enhancement with greater manual dexterity or physical power has been the subject not just of popular entertainment, but of actual bioengineering research that seems increasingly plausible as we take into account emerging capacities for information control—our new digital functioning. If a person can manipulate an artificial device, then those same mechanisms of control might well be applied to a machine of greater scope. Where then does the person end and the machine begin? Or is it not even important to be able to distinguish ourselves from the machines we make? Steve Mann, a professor in Toronto, has worn a computer for nearly twenty years, and has expressed the view that "wearable computers are not just a new kind of gadget, but the beginning of a fundamental shift in the relationship between people and technology." He sees his computer not as a tool, but as "an extension of his perception, his memory, and his identity."[25]

Rapid progress is also being made in understanding the neurochemistry of how the brain operates as an information processing unit. We are at the very beginning of learning how the brain manages information, enabling us to influence the external world. Some people believe that we will eventually achieve a level of understanding of brain functioning that will allow volitional control of external devices – that is, causing events to occur as we wish just by willing that it be so, rather than by

[23] Rossi.

[24] CNN London, "Face transplants not just science fiction."

[25] Young, A31-32.

manipulating mechanical control devices. Combine that with the example of enhanced prosthetics, as some have proposed, and we can envision hybrid human-machine collaborations that make it even harder to define what the limits are of the human as distinct from the machine. Do not think this is just the stuff of science fiction, where it has long been familiar. British Telecom actually initiated a research program named "Soul Catcher," which sought to develop an implantable chip that would enhance human memory and human calculating ability, and that would receive information by wireless networking.[26]

Susantha Goonatilike, in his recent book *Merged Evolution: Long-Term Implications of Biotechnology and Information Technology,* notes that biological, cultural, and artifactual lineages each takes on an evolutionary life of its own and that these lineages are quickly merging. He has envisioned the possibility of affirming a cultural value—appreciation for music—by implanting a device in humans that will allow extending their audible frequency range, thus using an artifact to enhance the biological organism's culture-capturing capacity. Such enhancement might also occur purely by genetic manipulation, using, for example, genetic material from species that hear more acutely than we do, and thereby eliminating the need for this new variety of silicone implant.

And what are the long-term consequences of functioning in an electronic environment? How does this technology influence our social interactions? What does this portend for the social dimensions of human nature? It is central to human existence that we do not live in isolation, but in various relationships to other people, in families, in communities, in the polis. As those modes of interpersonal relationships change in the context of cyberspace, the very nature of the human participants is subject to change.

VII.

Involvement in these issues, and therefore reflection and deliberation about them, will become unavoidable for many of us as we encounter new technologies in our personal lives. But I hope we will aspire to and insist on much more. Although it is not easy to explain fully and conclusively what the grounds are for our deep sense of obligation to human nature and the future of humans and the planet, the power and pervasiveness of that sense is compelling. In considering our future, we know we must struggle to see clearly what is best, to fight for that, and to protect against deterioration. As we consider our evolutionary future, we should understand that the stakes are just as high and the need for discernment, attentive listening, reasoned deliberation, and socially responsible advocacy are just as great. If enough of us do so, and if we proceed honestly and carefully with more humility than hubris, then

[26] Kotler.

human nature—enriched by the fruits of biomedical research—may have a future worth having after all, at least for a while.

Acknowledgements: I thank Melissa Freeman for excellent assistance in the preparation of citations and Christiana Peppard for superb editorial suggestions.

REFERENCES

Altman, Lawrence K. "Health panel seeks sweeping changes in fertility therapy." *New York Times,* late edition (Wednesday, April 29, 1998): A1.

CNN London. "Face transplants not just science fiction." November 28, 2002. http://www.cnn.com/2002/HEALTH/11/28/face.transplants/.

Dobzhansky, Theodosius. *Heredity and the Nature of Man.* New York: Harcourt, Brace and World, Inc., 1964.

Goodstein, David. Personal Communication. July 8, 2004.

Goonatilike, Susantha. *Merged Evolution: Long-Term Implications of Biotechnology and Information Technology.* World Futures General Evolution Series, vol. 15. Amsterdam: Gordon and Breach, 1999.

Huxley, Sir Julian. "The Future of Man: Evolutionary Aspects." Lecture at "Man and His Future" conference, November 27-30, 1962. Published in *Man and His Future.* Edited by Gordon Wolstenholme, 20-22. Boston: Little, Brown and Company, 1963.

Kass, Leon. "Beyond Biology." Review of *Brave New* Worlds, by Brian Appleyard. *New York Times Book Review* (August 23, 1998).

Kessler, David. *A Question of Intent.* New York: Perseus Books, 2001.

Kotler, Steven. "Tech 2010: #32 Everlasting: The Genius Who Sticks Around Forever." *New York Times* (June 11, 2000): Section 6: 102.

Maddox, Brenda. *Rosalind Franklin: The Dark Lady of DNA.* New York: Harper Collins, 2002.

Melville, Herman. Chapter 49: "The Hyena." In *Moby Dick* (Reissue Edition). New York: Bantam Books, 1981.

New York State Task Force on Life and the Law. *When Death is Sought: Assisted Suicide and Euthanasia in the Medical Context.* New York: May 1994.

New York State Task Force on Life and the Law. *Assisted Reproductive Technologies: Analysis and Recommendations for Public Policy.* New York: April 1998.

Plato, *The Republic.* Translated by G.M.A. Grube, revised by C.D.C. Reeve. Indianapolis: Hackett Publishing Company, 1992.

Pope, Alexander. *Essay on Man* (1733-1734). Epistle II, Line 1.

President's Council on Bioethics. *Human Cloning and Human Dignity: The Report of the President's Council on Bioethics.* Washington: PublicAffairs, 2002.

Regalado, Antonio. "The Troubled Hunt for the Ultimate Cell." *Technology Review* 101 (July/ Aug 1998): 34-41.

Rossi, Lisa. "Press Release: More hand transplants to be performed, predict experts at international congress of the transplantation society." August 26, 2002. http://www.txmiami2002.com/press7.htm.

Sartre, Jean-Paul. *Existentialism and the Human Emotions.* New York: The Wisdom Library, 1957.

United Nations Educational, Scientific and Cultural Organization. General Conference, 29th Session. *Universal Declaration on the Human Genome and Human Rights.* Paris, France: November 1997.

Varmus, Harold. Statement on Stem Cells to Senate Appropriations Subcommittee on Labor, Health and Human Services, Education and Related Agencies. January 26, 1999.

Vogel, Gretchen. "Breakthrough of the Year: Capturing the Promise of Youth." *Science Magazine* 286 (5448): 2238-9.

Wade, Nicholas. "Researchers claim embryonic cell mix of human and cow." *New York Times.* November 12, 1998: A-1.

"With Help of Science, Infertile Couples Can Even Pick Traits." *New York Times.* November 23, 1997.

Young, Jeffrey. "Self-Described 'Cyborg' Reveals Promise and Dangers of Wearable Computers." *The Chronicle of Higher Education* XLVIII, no. 34 (May 3, 2002): A31-32.

DAVID EHRENFELD

UNETHICAL CONTEXTS FOR ETHICAL QUESTIONS[1]

Like most biologists, I have suffered from occasional bouts of physics envy. One of the high points of my years at college occurred midway through a course in general physics—the only physics course I've ever taken after high school. We were studying sound, and somehow, probably using neurons that have long since died, I managed to derive without help the equation for the Doppler Effect. Suddenly a window opened up somewhere in my mind, and peering through I caught a glimpse of the glorious, crystalline world of theoretical physics, inviting me in. But before I could crawl through it, the window slammed shut, smashing a few more neurons in the process. Since then, I have had to content myself with the study of biology, not even molecular biology and biochemistry (apart from a few clumsy forays in medical school and graduate school), but physiology, ecology, behavior, natural history, and—I hesitate to admit it—a decades-long interest in the relationship between society, technology, and environment.

In the beginning of my academic career, the physics envy was stronger, and on the blackest days I found myself doubting whether Charles Darwin was as bright as Isaac Newton. The theories of physics appeared more comprehensive, more general, more fundamental, more mathematical, and more predictive than most of those of biology or any other field inside or outside science. However, as I got older and learned a little more about how the world works, I began to realize that physics envy was a waste of emotional energy. Physics actually deals with some of the simplest systems the human mind can comprehend: one or two sub-atomic particles, for example, or the statistical uniformity of the molecules of an ideal gas. The rest of us, meanwhile, cope with generality-defying, prediction-defying complexity, especially as we stray farther and farther from physics, even as far afield as ethics. We all long for simplicity, but we are bogged down in a sea of interacting variables. It doesn't seem fair.

Looking a little more closely, however, I began to see that physics is not really simple. The dynamic chaos of earth, sky, moon, planets, and stars that Newton saw around him did not resolve itself into laws of motion, gravitation, and cooling. He had to find the simplicities, the basic patterns in the realm of creation he chose to study; he had to find the right explanatory contexts in which his genius could operate effectively. Contemporary physicists are faced with the same problem, and it has become clear that the models for and solutions to "basic" questions do not help answer those of a less basic nature; other contexts are required. The models of elementary particle physics are of little use in helping physicists understand the less

[1] A modified version of this essay was published under the title, "The Cow Tipping Point" in *Harper's Magazine* 305 (October 2002): 13-20.

A. W. Galston and C. Z. Peppard (eds.), Expanding Horizons in Bioethics, 19-34.

fundamental solid state or many-body physics (although the converse is not necessarily true). As the distinguished physicist P.W. Anderson wrote in a celebrated paper:

> The reductionist hypothesis does not by any means imply a "constructionist" one: The ability to reduce everything to simple fundamental iaws does not imply the ability to start from those laws and reconstruct the universe. In fact, the more the elementary particle physicists tell us about the nature of the fundamental laws, the less relevance they seem to have to the very real problems of the rest of science, much less to those of society.[2]

Darwin had the same task of finding the right context: at home and in his travels he encountered many settings of nature, many contexts, and profited from them all; but not until he reached the Galapagos Islands did he come across the context for understanding evolution. His special gift, like that of Newton, was recognizing it.

Conversely, choosing the wrong context when answering a particular question, and especially confining oneself to an overly narrow context, is very likely to lead one astray. For example, the isolated gene has been a useful context for answering some of biology's most important questions, but for other questions, as the Harvard biologist Richard Lewontin has pointed out, the gene alone, considered apart from the intracellular, extracellular, and external environments in which it functions, has proven an unreliable context for answering questions such as, What is the cause of schizophrenia?[3] 40 years ago, another Harvard biologist, Ernst Mayr, made similar observations about the limitations of isolated single genes as a context. In a chapter entitled "The Unity of the Genotype" in his book *Animal Species and Evolution*, he made compelling arguments to show that the effect of genes was strongly modified by other genes and by the internal and external environments, and varied from individual to individual.[4] This was based on powerful research findings in studies going as far back as 1937, and earlier. Unfortunately, many of today's genetic engineers give no sign that they have heard the message of Lewontin and Mayr.[5]

The matter of context also applies to ethical questions. How can we decide what is right and what is wrong, what to do or not to do in these times, which are probably the most complex in human history because we have taken on the management of so many more things in the world than ever before? In my book, *Swimming Lessons: Keeping Afloat in the Age of Technology*, I discuss the words of the 18th century Ukrainian philosopher, Gregory Skovoroda, who said: “We must be grateful to God that he created the world in such a way that everything simple is true, and everything complicated is untrue.”[6] I wish I could believe this, but I am afraid that I do not. Nevertheless, there are ways to cut through the complexity, to approach the underlying simplicities, the basic truths that we must see if we are to survive as individuals and as a society in this Age of Management. This is what modern giants such as Lewis Mumford, George Orwell, Heinrich Böll, Jane Jacobs, Mary Midgley,

[2] Anderson.
[3] Lewontin.
[4] Mayr.
[5] Cf. Commoner.
[6] Ehrenfeld, *Swimming Lessons*.

Wendell Berry, and Vaclav Havel have done, each finding the right contexts—usually very spacious ones—in which to set their questions. But to find the right contexts, we must survey as many of them as possible. If we restrict ourselves to a narrow context, complexity is very likely to hide the basic truths.

This is not a problem for everybody, however. There are those who, for one reason or another and usually for short-term personal advantage, do not care to approach the truth too closely. For them, complexity is a godsend. Like a squid escaping its pursuers in a cloud of ink, they can use complexity to obscure their movements, to hide the significance of what they are doing. By selecting the narrowest from the many available contexts in which to portray and evaluate their own actions, and by cloaking these actions in a haze of technological intricacy, people can get away with behavior that society would not countenance if it were thinking clearly. I will give some concrete examples.

Recombinant bovine growth hormone, rBGH for short, sometimes called recombinant bovine somatotropin, or rBST, is a growth hormone for cattle produced by taking the growth hormone gene from cows, modifying it very slightly, and inserting it into bacteria, using techniques of genetic engineering.[7] The altered *E. coli* bacteria can be grown in vats, producing large quantities of rBGH, vastly more than could be obtained economically by extracting the unmodified growth hormone directly from cows. This rBGH, like its parent gene, is very slightly different from the natural product, having a substitution of just one amino acid for another at the end of the large molecule. In the United States, rBGH is marketed by Monsanto under the name of Posilac. When injected into lactating cows, it increases overall milk yields by approximately 10-15%, although greater increases are occasionally observed.[8]

This is a dramatic kind of biotechnology, albeit dependent on a relatively rare phenomenon: a single gene coding for a product that is directly or indirectly commercially valuable. Not surprisingly, as is the case with all new technologies that cause radical changes in production systems, economics, and cultural systems, the marketing of rBGH has engendered a great deal of controversy.

From the beginning, the controversy swirled around two questions: Is the milk from cows injected with rBGH different from milk from untreated cows; and if so, is it harmful to the humans who drink it? Second: Does the injection of rBGH into lactating cows harm the animals in any way? Monsanto has not been able to provide an unequivocal no to either of these questions, and this may be part of the reason why Posilac has, by many accounts, not proven to be a cash cow for the company. Yet I imagine (I cannot prove it) that Monsanto would prefer to keep the rBGH controversy confined to these issues, because the context of the questions is pleasingly narrow—in other words, most of the ethical concerns generated by the use of rBGH do not come up at all. Moreover, the two questions, because of their nature, can be drawn into a mire of complex and often contradictory technical and scientific details that make clear judgments difficult to achieve. This confusion works well for Monsanto, because the company wants sales, not judgments.

[7] Hansen.
[8] Coghlan.

Before widening the scope of the inquiry, I want to dispose of the two questions. Is rBGH milk different from other milk? Yes and No. According to a paper published by Samuel Epstein in the *International Journal of Health Services* in 1996,[9] and earlier reports summarized by T. B. Mepham in the *Journal of the Royal Society of Medicine* in 1992,[10] rBGH milk contains elevated levels of Insulin-like Growth Factor-1 (IGF-1), a suspected cause of human breast and gastrointestinal cancers. Supporters of rBGH are quick to point out that IGF-1 also occurs in milk from untreated cows; and that its carcinogenic effect is not conclusively proven. Opponents respond that there is at least a three- or four-fold increase of IGF-1 in rBGH milk, and that more of it may be in an unbound, free form, which might be biologically more active. It also should be noted that rBGH itself is present in the milk of treated cows, perhaps in elevated levels over the natural hormone, and it is possible that this unnatural protein could cause allergic reactions or, after partial digestion in the human gut, mimic the metabolic effects of human growth hormone. Lots of "mights" and "maybes," credible suspicion but no proof, no smoking gun. The ink is swirling in clouds. Let's look at the second question: Does rBGH injection harm cows?

At first glance, rBGH does not come off so well. To avoid charges of anti-rBGH bias, I will take my information from the package insert for Posilac ("sterile sometribove zinc suspension"), copyright by Monsanto in 1993. According to the manufacturer's label, use of Posilac causes "feed intake increases over several weeks" after starting injections. No surprise there; the laws of thermodynamics hold for cows. The animals are producing more milk, so they must eat more food—I will come back to this later. Use of Posilac also "may result in reduced pregnancy rates... increases in cystic ovaries and disorders of the uterus...small decreases in gestation length and birth weight of calves... reductions in hemoglobin and hematocrit values... periods of increased body temperature unrelated to illness... indigestion, bloat, and diarrhea... increased numbers of enlarged hocks and lesions [of the knee]... [and] disorders of the foot region." But the biggest health problem for rBGH-injected cows is "an increased risk for clinical mastitis (visibly abnormal milk).... In addition, the risk of subclinical mastitis (milk not visibly abnormal) is increased." "Visibly abnormal milk" means pus in the milk.

The label's recommendations for how to cope with this constellation of problems seem quite sensible. I will condense and paraphrase them: Be sure you are ready to deal with increased veterinary problems, presumably by keeping more veterinarians on staff or on call; be ready to differentiate between fevers caused by rBGH and fevers caused by disease; and for cows running a fever, control heat stress "during periods of high environmental temperature," I suppose by means of air conditioning; and, implement a "comprehensive and ongoing herd reproductive health program," whatever that means.

It is worth noting that none of the ailments listed as being associated with rBGH injection are unique to this treatment; cows can get mastitis, bloat, and sore knees even if they are raised under strict conditions of organic husbandry. And Monsanto

[9] Epstein.
[10] Mepham.

has pointed out that the increase in mastitis may be a result of increased milk production itself, thus only indirectly caused by rBGH injection. The clouds of ink thicken. Again, we are left with legitimate worries that have not been properly addressed by the Food and Drug Administration (FDA), but without an absolutely clear-cut mandate to condemn the technology.

In a situation of this kind, what usually happens is a continuation of the status quo. The results of peer-reviewed research produced by independent scientists are contradicted by the results of peer-reviewed research sponsored by the company. Each study, regardless of authorship, is run in a different way under different conditions, making comparisons problematic. Some necessary analyses, such as distinguishing between natural BGH and rBGH in milk, prove difficult or impossible. The federal regulators, some of whom were formerly executives in the regulated industry, feel justified in keeping the product on the market. And the worries persist.

This is the time to widen the context of the inquiry, to reject efforts to keep questions confined to a narrow space where visibility can always be obscured by more convenient ink. I propose to widen the context gradually so that we always know the vantage point from which we are viewing the bioethical landscape. Eventually, the basic truths of the matter should be fairly clear, if they aren't already; and the conclusions we ought to reach about the technology will be obvious.

The first small step to take is to see what happens when we merge questions one and two. The most solid finding from the inquiry into the effects of rBGH on the health of cows is that treated cows get significantly more mastitis than untreated ones. This is a finding admitted by Monsanto and confirmed by the FDA.[11] Mastitis in cows, like breast infections in humans, is usually treated with antibiotics, and these antibiotics may well find their way into the milk. In an ideal world, milk containing antibiotics is kept off the market. This is not an ideal world. Government agencies test milk for only a small number of antibiotics, and they do not test every batch; there are many antibiotics that can slip through into supermarket milk. Careless or unscrupulous farmers may sell milk containing antibiotics, and some farmers may be willing to deliberately treat their cows with antibiotics that they know are not going to be screened in government tests. When antibiotics get into the milk, antibiotic resistance can be transferred from the bacteria normally in the milk to the bacteria that normally live in the intestinal tract of humans, and this resistance can be transferred again during illness to the bacteria causing the disease.[12] The result is that when antibiotics are given to sick people, they do not work.

An analogous case concerns the refusal, early in 2001, of the Bayer Corporation to heed the request of the Centers for Disease Control and Prevention and the Food and Drug Administration's Center for Veterinary Medicine to stop selling their fluoroquinolone antibiotics for routine use as growth promoters in the diets of factory-farmed chickens, turkeys, and pigs. There is mounting evidence that feeding fluoroquinolones to poultry is causing widespread bacterial antibiotic resistance that

[11] Coghlan.

[12] Ferber, "Superbugs on the hoof?"

can make these drugs useless for treating many human infections.[13] (The most familiar of the fluoroquinolones is ciprofloxacin, sold under the brand name of Cipro™.)

Let's widen the context a little more. I mentioned earlier that rBGH injection increases the food intake of cattle; they need more calories, particularly in the form of protein. One of the best and cheapest sources of high-grade protein is the carcasses of dead farm animals, including sheep, horses, cows, and others. For at least 100 years, the rendering industry has been converting dead animals into food supplements for livestock, but the advent of high-milk-yielding cattle and, especially, rBGH-injected cattle, has greatly increased the demand for this animal protein in cow fodder. Cows have been turned into carnivores, even cannibals. In recent years, we have become aware, however, that a terrible neurological disease, worse than Alzheimer's, called spongiform encephalopathy, is transmitted from individual to individual and even from species to species by eating brain, nerve, and other tissue from infected animals. In cattle, we call this mad cow disease; in deer and elk chronic wasting disease; in humans it is Creutzfeldt-Jakob disease; and there is little doubt that it has been spread in England and the Continent by the practice of feeding rendered, processed carcasses of other ruminants to cattle.[14] Here, then, is another legitimate and serious worry caused by the use of rBGH: will it increase the incidence of spongiform encephalopathy/mad cow disease/Creutzfeldt-Jakob disease in the United States, where this constellation of diseases already exists?

As we move farther and farther from the original narrow context, we gradually leave the realm of science and medicine and we enter the territory of ethics, economics, and social well-being. Our next consideration in this widening inquiry takes us to the welfare of cattle. Even if we ignore the ethical implications of increased disease caused by rBGH, there are other important questions to be considered. Do we have the right to treat cows as if they were mere machines for producing milk, with all the suffering and lack of respect that this implies? Do we have the right to burn them out, to shorten their useful and productive lives, which is what rBGH appears to do? According to the farmer and agricultural writer Gene Logsdon, dairy farmers used to be able to keep their cows on the milking line for twelve to fifteen years; now, with many cows being treated with rBGH, they frequently last only two or three years.[15] Accordingly, the price of replacement heifers has risen sharply, reflecting the increased demand.

Now we can widen the context again and look at the welfare and rights of dairy farmers, and, beyond that, at the welfare of the communities and larger society in which they live. Matthew Shulman, owner of a small farm in Lansing, New York, and former director of information for the New York State Grange and executive secretary of the New York State Forage and Grasslands Council, was one of the first to write on this subject.[16] He questioned the claim of proponents of rBGH that this technology is farm-neutral, that if used properly it will work as well on small farms

[13] Falkow and Kennedy, 1390.
[14] Pattison.
[15] Logsdon, personal communication.
[16] Shulman.

as on large ones. He was concerned with the prohibitive cost of high-tech feed management systems and high-protein rations, which would price rBGH right out of the market for small farmers. He also noted that the hormone was marketed primarily to large farms, anyway. Shulman's argument would have been even stronger if he had known more about the increased veterinary, air conditioning, and cow replacement costs associated with the use of rbGH. All of these costs can only be borne by large farming operations, which typically carry much higher levels of debt than small farms.

Four years later, Charles Geisler and Thomas Lyson, professors of rural sociology at Cornell, confirmed Shulman's fears in an article on the social and environmental costs of dairy farm industrialization. As Geisler and Lyson pointed out, large dairy farms have: lower technological diversity, a higher rate of accidents, worse environmental impacts, increased dependence on specialized wage labor with lower system resilience (manifested as an increased likelihood of strikes), decreased personal knowledge of individual animals, and, finally, greater centralized control and more non-resident owners, with a consequent breakdown in "economic vitality and social cohesion in rural communities."[17] A big part of the problem, they wrote, is debt; farm debt as a percentage of a farm's value increases dramatically as the size of its dairy herd increases. And as the debt-to-asset ratio increases, partly to pay for the supplementary, expensive veterinary care, climate control for feverish cows, and high-priced feed supplements that go along with the use of rBGH, control of dairy farming shifts away from the farmer and the farm community to distant banks. Then, Geisler and Lyson state, "as debt continues to rise, the dairy industry will be increasingly sensitive to non-local production factors, such as... interest costs."[18]

Once the small dairy farms are gone, the industrialized farms that remain will become completely dependent on the new milk production technologies because they cannot produce milk any other way. This will lead to the same kind of corporate vertical integration that has placed a few oil, chemical, and pharmaceutical companies in control of much of the world's agricultural seed production, resulting in the rapid, irreversible loss of thousands of agricultural food varieties of great and irreplaceable value, and putting the world's food supply in jeopardy.[19]

There is one more context in which I want to evaluate rBGH. In the eastern states from North Carolina to Massachusetts, and beyond, small dairy farms have long given a particular look and character to the rural countryside. Typically, such a farm comprises 80-95% upland pasture and 5-20% wet grazing areas: stream corridors, marshes, and bogs. The whole is divided into small fields through which the cattle are rotated. It has become clear in recent years that the cows on these small dairy farms accomplish much more than just milk production. They have serendipitously replaced, in the wetland areas, other large eastern grazing mammals, the mastodons, elk, and bison, which have been progressively eliminated by waves of human settlers, starting eleven or twelve thousand years ago. Like these former native grazers, cows eat and therefore control the invasive, and these days often

[17] Geisler and Lyson.
[18] Ibid.
[19] Fowler and Mooney.

exotic, species that are modifying and eliminating wetland species and plant communities. They eat red maples and alders, *Phragmites*, reed canary grass, purple loosestrife, and similar invasives that otherwise choke out wetland vegetation all over these eastern states. Thus, if you want to find the tiny bog turtle (*Clemmys muhlenbergii*), the fen buck moth (*Hemileuca sp. 2*), the showy lady slipper orchid (*Cyripedium reginae*), or the spreading globeflower (*Trollius l. laxus*)—all of them rare and endangered—you will have to go to a small dairy farm, or land that was a small dairy farm until recently; you probably will not find them anywhere else.[20] So here is yet another effect of rBGH: the big, industrialized dairy farms that rBGH promotes, with cows being fed high-protein food supplements in temperature-controlled buildings, do not serve the smaller farms' unexpected function of maintaining the flora and fauna of wetlands.

Why has the United States, which in 1986 and 1987 paid 14,000 dairy farmers $1.8 billion to slaughter 1.55 million dairy cows to reduce the milk glut, and which between 1987 and 1989 paid between $600 million and $1.3 billion a year to purchase surplus milk, been pushing rBGH so hard? And how has the government gotten away with it? The first question is easy to answer: Monsanto and similar companies have been major contributors to both the Republican and Democratic parties. The second question is easy to answer, too. The government has gotten away with it because it has confined the ethical debate to the narrowest possible context, where the waters were muddy and the larger issues lay hidden.

Even with this context restriction, the case for rBGH is so weak that only the most skillful political damage control has kept it on the market in the U.S. Canada and the European Union have both banned rBGH, on the significant but narrow grounds of animal health. And in 1999, the Codex Alimentarius, the food safety standard organization of the Food and Agricultural Organization and the World Health Organization of the United Nations, refused to certify rBGH as safe. It effectively tabled the rBGH issue as a way of saving face for the U.S., which would have lost a formal vote.

In summary, we must look at the entire picture of the effects of rBGH: this is not only IGF-1 in the milk and animal health, but antibiotic resistance, spongiform encephalopathy, animal welfare, the welfare of farmers and farm communities, the well-being of agriculture, and the maintenance of whole ecosystems. Is it legitimate to widen the context so broadly in evaluating a new technology? Yes; it is more than legitimate. It is practically and ethically essential if the the truth is to emerge, for the message produced by these overlapping and widened contexts is really quite simple to understand: rBGH is a very bad technology indeed.

Having examined the rBGH controversy in some detail, it might be instructive to look more briefly at a few other examples that show the value of contextual widening. Genetically modified food (GM food) is a category somewhat different from that of rBGH milk. "GM food" is food made from crops that have received genes via genetic transfer from other organisms, even distantly related ones. Salmon, for example, can be engineered to contain human genes, and corn now contains bacterial genes. Most of the GM food on the market is from crops either

[20] Lee and Norden; Tesauro.

engineered to produce an insecticidal bacterial polypeptide commonly known as the Bt toxin, or to produce enzymes that inactivate the seed company's brand of weed killer, as in the case of rapeseed and soybeans.[21] In the former instance, crops producing their own Bt toxin are marketed to farmers as requiring less external insecticide application. In the latter case, conversely, farmers are told that they can liberally apply the company's herbicide without fear of damaging their crops.

Again, the proponents of GM foods have tried to keep the evaluative context as narrow as possible, asking: Do these foods contain harmful substances? And again, apart from a few obvious mistakes involving genes from highly allergenic foods such as brazil nuts transferred to soybeans, and genetically engineered gene products not approved for human consumption introduced into corn, the question of toxicity does not yield a clear answer; there is a lot of scientific-technical ink in the water. The anti-GM group notes that these foods contain alien polypeptides and proteins, which might cause illness in susceptible individuals. The pro-GM voices respond that plants have been producing toxic chemicals to kill insects and competing plants for millions of years: witness the insecticide nicotine in tobacco, and the ghastly compound produced by tobacco's cousin, the white potato, if you expose the growing tubers to sunlight. True, reply the antis, but we have had millions of years to evolve biological and cultural responses to natural toxins in the food we eat. Fine, retort the pros, but what about this: nature was moving genes between species for countless millennia before agriculture began; and, further, conventional plant and animal breeding, which everyone accepts, also shuffles genes from one variety to another—even from one species to another. Yes, respond the antis, but not between spiders and goats, or people and pigs. And so it goes. It is time to widen the context.

For GM foods, much more than rBGH, there has already been a little context widening. Newspapers have documented the probably deleterious effect of Bt-containing crops on monarch and other butterflies, and some public comment has emerged about the general damage to pollinating insects from continuous exposure to the insecticide produced by GM crops day after day, month after month, over hundreds of thousands or millions of acres. Public mention has even been made of the fact that GM crops can move herbicide resistance genes into the weeds and cause insect pests to evolve resistance to the Bt toxin.[22] A less publicized facet of the Bt story is that the loss of effectiveness of this natural insecticide could put many organic farmers out of business, a side effect that might not displease the chemical companies that own the seed companies that make the GM crops.

When we widen the context, however, to include the well-being of the farmers using GM crops, public attention drops off. In the late 1990s there was some brief publicity given to cotton farmers in the south, who brought suit against the manufacturer of GM crops because the crops, they claimed, did not work as advertised, and because the expensive, one-time technology use agreements they were required to purchase with the GM seed were only good for one planting. But until recently I have seen comparatively little public mention of the case of the

[21] Teitel and Wilson; Ho.

[22] Holmes, 7; Robert and Baumann; Ferber, "New corn plant draws fire," 1390.

Canadian farmer, Percy Schmeiser, whose rapeseed crop was discovered by Monsanto's "gene police" to contain patented genes for herbicide resistance from Monsanto's GM rapeseed, which Schmeiser had never purchased.[23] As has been repeatedly demonstrated, pollen containing industrially produced GM genes, can move considerable distances to enter both conventional crops and wild plants.[24, 25] Despite the possibility that the patented GM rapeseed genes got into Schmeiser's crop not by deliberate theft but by pollen blown from the GM rapeseed fields across the street from his farm, and despite the fact that Schmeiser claimed he was not using the Monsanto herbicide that would have let him benefit from those patented genes, a Canadian lower court judge found Schmeiser guilty and ordered him to pay Monsanto heavy damages.[26] Schmeiser's case is not unique. Other farmers in the United States and Canada have been assessed damages by Monsanto for alleged infringement of gene patents. The implications are chilling for farmers everywhere who don't want to buy genetically modified crop seeds. Why isn't this context receiving careful scrutiny?

Clearly, beyond the narrow question of whether GM food will make you sick, there are enormous problems with the technology—and we are not looking at those questions. I will briefly mention three more. First, there seems little doubt that the introduction and widespread use of GM food crops will cause a further reduction in the number of major crop varieties in existence, a process already started by the industrialization of agriculture. This will narrow the genetic base of agriculture, which in turn will paradoxically limit the future opportunities for both genetic engineering and conventional breeding. Reduction in the number of crop varieties will also make us more vulnerable to the spread of crop pathogens by terrorists, a fact well-known to the bioterrorism taskforce.

Second, and related to the first, the granting of industrial patents for genetically modified crops (including crops whose genes have been sequenced but hardly modified at all) allows a handful of corporations to own and control much of the world's food. This development seems at least as worthy of discussion as the current activities of terrorists. I am convinced that profitability has little to do with the corporate push to introduce GM crops. They often don't work very well: they do not necessarily increase yields, nor will they outlive the development of insecticide and herbicide resistance in insect pests and weeds. Sooner or later their sales will slump, and the industry knows it. The real reason for this technology is that it opens the door to the corporate patenting and ownership of our food crops.[27]

Third, there is one more context which is of special importance to those Jews, Muslims, Hindus, and others who observe ritual purity laws for food. Is the food

[23] Rural Advancement Foundation International.

[24] Klinger, Elam and Ellstrand; Ellstrand; Quist and Chapela.

[25] See also the debates about the paper by Quist and Chapela, which were published in *Nature* 416 (2002): 600-602; *Nature* 417 (2002): 897-8; and Mann.

[26] A court later held that Schmeiser did not have to pay damages or legal fees to Monsanto because he did not intend to "steal" the seed.

[27] Hobbelink.

acceptable if it contains gene products from unacceptable species? Most religious authorities have not begun to deal with this problem.

Therefore, for all the reasons I have mentioned, I believe it is unethical to confine the GM food issue to a discussion of direct toxicity. Yet this is exactly what has been done. For example, the medical ethicist Marc Lappé, who was chosen as a consultant for a National Research Council study of a major group of genetically engineered crops, "was asked to limit my scrutiny to scientific data and to focus solely on scientific questions of risk while eschewing political, social, or philosophical issues." Therefore, "the NRC asked for a review that had a foreordained answer: Insufficient evidence exists on which to base health concerns from GMOs."[28]

Not surprisingly, problems of context extend beyond the realm of agriculture and genetically modified foods. My next example of a context problem in bioethics is that of human reproductive (and possibly therapeutic) cloning.[29] I will give it short shrift, however, because I agree with M.I.T.'s Rudolf Jaenisch and the Roslin Institute's Ian Wilmut, cloner of Dolly, that human cloning is not going to work in an acceptable way for the forseeable future.[30] Also, if reproductive cloning can ever be made safe and reliable, a very big *if*, the recipients of cloned children (those who are not just growing them for spare parts) will be terribly disappointed, because cloning—for biological, cultural, and environmental reasons—does not produce xerographic copies.

"That's not me!" I can imagine an angry, Fortune 500 CEO shouting, as he bails his shiftless, stupid, cloned, adolescent son out of jail. "He doesn't even look like me!" Enter the lawyers.

But questions of feasibility do not stop us from talking about human cloning, so I will mention just one newspaper article entitled "Two Cheers For Human Cloning," by Sheryl Gay Stolberg.[31] It is a good article, examining many of the moral and some of the technical arguments pro and con. It states them very clearly, repeating the points of view of scientists, ethicists, potential parents of cloned children, Congress, and even the biotechnology industry (which appears to oppose human cloning for the strategic reason of preserving its related research). But nowhere in the article is there any discussion of the context outside that particular ethical box: namely, what will happen, ethically and practically, should human cloning become commercially available? What will be the consequences of granting such immense power over human reproduction to a few corporations and non-profit organizations?

My last example of the value and ethical necessity of context widening comes not from biotechnology but from the practice of species conservation. A 300-square-mile chunk of the Edwards Plateau of central Texas is occupied by the U.S. military base called Fort Hood. On that base, as elsewhere on the plateau, live two endangered species of birds: the black-capped vireo and the golden-cheeked warbler,

[28] Lappé, "A perspective on anti-biotechnology convictions."

[29] Human reproductive cloning, which aims to produce a human person, should be distinguished from therapeutic cloning, in which a primary goal is to glean useable stem cells and tissues for research.

[30] Jaenisch and Wilmut.

[31] Stolberg.

each of whose total remaining populations number in the hundreds or low thousands of individuals. Although protected on the base, these two species were in decline until recently because of nest parasitism by the brown cowbird. The cowbirds, which are anything but endangered, and which are also native residents of the plateau, always lay their eggs in the nests of other species of birds. The cowbird chicks hatch first and hog most of the food brought by their hapless foster parents. Few vireo and warbler chicks from cowbird-infested nests survive.

For the past few years, wildlife managers at Fort Hood, in cooperation with The Nature Conservancy, have been trapping cowbirds in gigantic cages baited with grain, then killing the adult females. Predictably, vireo and warbler populations have increased on the base in response. So far, the minimal ethical debates about this kind of conservation technique have centered entirely on the rights of warblers and vireos, the rights of cowbirds, and the rights of people opposed to the killing of animals. You can imagine the ethical quagmire here. There are no simple answers.

So let us widen the context with a question: Why are cowbirds threatening vireos and warblers now, when before the past century they coexisted for countless millennia? The answer is ecological. Cowbirds feed in the open grasslands; vireos and warblers nest in adjacent brushy shrublands. The cowbirds parasitize nests built near edge areas where grassland and shrubland meet, not in the interior parts of the shrublands. Uncontrolled human land use on the base, particularly conversion of some of the hilly shrublands to grasslands for grazing by cattle, and—more importantly—unplanned, minimally regulated suburban sprawl off-base around the cities of Austin, San Antonio, and Waco, have fragmented and dissected the shrubland to the point where it is mostly edge habitat now. So the cowbird is not the real culprit; it is merely the last straw. Nevertheless, it is far more convenient and politically expedient to blame cowbirds rather than limit the cattle grazing on the base or make an effort to introduce responsible zoning and land management around the cities, while phasing out the cowbird extermination program.[32]

With so many ethical stalemates occurring in agricultural and reproductive biotechnology, and even in conservation, why do we fail to widen the context when we debate these critical issues affecting society? It is not just because we are being kept to a narrow, controllable venue of debate by vested interests, although that is usually the case. Nor is it just that much of the public, dumbed and numbed by television and advertising, is incapable of digesting anything more complicated than a sound bite. I think the deeper problem is that more than 200 years of potent scientific discoveries and technological inventions—from the steam engine to the laser scalpel—have taught us to believe that science and technology, the fruits of our own reason, constitute the highest power we need consult in our daily lives. In our euphoria we forget two things. First, technology is unable, both in theory and in practice, to resolve all of the practical problems that it, itself, creates.[33] Second, science and the exercise of reason cannot by themselves provide the moral framework we need to judge our own inventions.

[32] Ehrenfeld, "Extinction and blame."
[33] Schwartz.

In my book, *The Arrogance of Humanism*, I said much the same thing twenty-four years ago: "Pure reason [does not] suffice to distinguish the humane and the just from the inhumane and the unjust."[34] John Ralston Saul, in *Voltaire's Bastards*, speaks of "reason's innate amorality."[35] That is, if we restrict the context of our ethical inquiries to a narrow review of scientific facts, if we respect only technical information, we may never reach the sources of wisdom best suited to guide us on a just and sustainable path.

Despite the many differences that divide us, human societies have achieved a remarkable consensus about what is right and what is wrong. In the eighteenth century, an extraordinary group of scientists, educators, reformers, and political leaders in Europe and America was able to find the basic truths in the complex affairs of national and personal life. Neil Postman describes these people in *Building a Bridge to the 18th Century*—not only Voltaire but Diderot, Paine, Franklin, Jefferson, Madison, and many others.[36] They were *philosophes*, not philosophers; they were interested in finding solutions to the practical, concrete problems of the day. All of them believed, according to Postman, that "answers to the question, What is the right thing to do? (or more precisely, What is the wrong thing to do?)" were obvious, based on a transcendental authority which all of them recognized, whether they called it God, Natural Law, Practical Reason, First Principles or Traditional Morality.[37] It was this ability to superimpose a common ethical structure on the complex problems they encountered that led all of them to conclude, as Postman notes, that "the Inquisition, slavery, debtors' prisons, torture, [and] tyranny" were wrong.

We can do the same today, as Wendell Berry counsels in *Life is a Miracle*, provided we do not expect to derive our moral law from our science. As Berry says, "Applying knowledge—scientific or otherwise—is an art."[38] I believe that the necessary practice of this art in the twenty-first century will not proceed until we refuse to limit the contexts of our inquiries. And only when this happens will ethics become more than, to quote the food policy expert, Brewster Kneen, "an enteric coating on the bitter pill being forced down the public throat."[39]

Of course there are no perfect answers to many of our ethical questions. Isaiah Berlin has reminded us that the "Great Goods can collide... human creativity may depend on a variety of mutually exclusive choices." When this happens,

> compromises can be reached.... Priorities, never final and absolute, must be established.... The concrete situation is almost everything. There is no escape; we must decide as we decide; moral risks cannot, at times, be avoided. All we can ask for is that none of the relevant factors be ignored.[40]

[34] Ehrenfeld, *The Arrogance of Humanism*, 81.
[35] Saul.
[36] Postman.
[37] Ibid.
[38] Berry.
[39] Kneen, 45-59.
[40] Berlin.

And that is all I ask for, that in our most important decisions no context be neglected.

It is up to us—as a society and as individuals—to frame our ethical questions properly. Ethicists should not do it for us; this is a process too important to leave to the professionals.[41] There is urgency for us to ask these questions of ourselves. The great contemporary British philosopher Mary Midgley has warned, "The house is on fire; we must wake up from this dream and do something about it."[42] But to do something appropriate we must make the right decisions. And these, in turn, require our taking the most inclusive view of the contexts of our activities that we can command in the time available.

Acknowledgements: I thank Dr. Naftaly Minsky for a stimulating discussion of reductionism in physics, and for directing my attention to the paper by P.W. Anderson. My wife, Dr. Joan Ehrenfeld, provided her usual helpful comments and suggestions.

REFERENCES

Anderson, P.W. "More is different: broken symmetry and the nature of the hierarchical structure of science." *Science* 177 (1972): 393-396.

Berlin, I. "The pursuit of the ideal." In *The Crooked Timber of Humanity: Chapters in the History of Ideas.* London: John Murray, 1990.

Berry, W. *Life is a Miracle: An Essay Against Modern Superstition.* Washington, D.C.: Counterpoint, 2000.

Commoner, B. "Unravelling the DNA myth: the spurious foundation of genetic engineering." *Harper's Magazine* (February 2002): 39-47.

Coghlan, A. "Arguing till the cows come home." *New Scientist* (29 October 1994): 14-15.

Ehrenfeld, D. *The Arrogance of Humanism.* New York: Oxford University Press, 1978.

________. "The Cow Tipping Point." *Harper's Magazine* 305 (October 2002): 13-20.

________. "Extinction and blame." *Orion* 20 (2001): 12-14.

[41] McKnight.
[42] Midgley.

_______. *Swimming Lessons: Keeping Afloat in the Age of Technology.* New York: Oxford University Press, 2002.

Ellstrand, N.C. "When transgenes wander, should we worry?" In *Engineering the Farm: Ethical and Social Aspects of Agricultural Biotechnology.* Edited by B. Bailey and M. Lappé, 61-66. Washington, D.C.: Island Press, 2002.

Epstein, S.S. "Unlabeled milk from cows treated with biosynthetic growth hormones: a case of regulatory abdication." *International Journal of Health Services* 26 (1996): 173-85.

Falkow, S. and D. Kennedy. "Antibiotics, animals, and people – again!" *Science* 291 (2001): 1390.

Ferber, D. "New corn plant draws fire from GM food opponents." *Science* 287 (2000): 1390.

_______. "Superbugs on the hoof?" *Science* 286 (2000): 792-4.

Fowler, C. and P. Mooney. *Shattering: Food, Politics and the Loss of Genetic Diversity.* Tucson: University of Arizona Press, 1990.

Geisler, C. and T. Lyson. "The cumulative impact of dairy industry restructuring." *BioScience* 41 (1991): 560-657.

Hansen, M. "Biotechnology and milk: benefit or threat? An analysis of issues related to bGH/bST use in the dairy industry." Mount Vernon, NY: Consumer Policy Institute/ Consumers Union, 1990.

Ho, Mae-Wan. *Genetic Engineering: Dream or Nightmare?*, revised edition. Dublin: Gill and MacMillan, 1999.

Hobbelink, H. *Biotechnology and the Future of World Agriculture.* London: Zed Books, 1991.

Holmes, B. "Caterpillar's revenge." *New Scientist* (6 December 1997).

Jaenisch, R. and I. Wilmut. "Don't clone humans!" *Science* 291 (2001): 2552.

Klinger, T., D.R. Elam, and N.C. Ellstrand. "Radish as a model system for the study of engineered gene escape rates via crop-weed mating." *Conservation Biology* 5 (1991): 531-5.

Kneen, B. "A naturalist looks at agricultural biotechnology." In *Engineering the Farm: Ethical and Social Aspects of Agricultural Biotechnology.* Edited by B. Bailey and M. Lappé, 45-59. Washington, D.C.: Island Press, 2002.

Lappé, M. "A perspective on anti-biotechnology convictions." In *Engineering the Farm: Ethical and Social Aspects of Agricultural Biotechnology.* Edited by B. Bailey and M. Lappé, 135-56. Washington, D.C.: Island Press, 2002.

Lee, D.S. and A.W. Norden. "The distribution, ecology and conservation needs of bog turtles, with special emphasis on Maryland." *Maryland Naturalist* 40 (1996): 7-46.

Lewontin, R. *It Ain't Necessarily So: The Dream of the Human Genome and Other Illusions.* New York: New York Review Books, 2000.

Logsdon, Gene. Personal communication.

Mayr, E. *Animal Species and Evolution.* Cambridge: Harvard University Press, 1963.

Mann, C.C. "Transgene data deemed unconvincing." *Science* 296 (2002): 236-7.

McKnight, J. *The Careless Society: Community and Its Counterfeits.* New York: Basic Books, 1995.

Mepham, T.B. "Public health implications of bovine somatotrophin use in dairying: discussion paper." *Journal of the Royal Society of Medicine* 85 (1992): 736-9.

Midgley, M. "Why smartness is not enough." In *Rethinking the Curriculum.* Edited by M.E. Clark and S. Wawrytko. Westport: Greenwood Press, 1990.

Pattison, Sir J. "The emergence of bovine spongiform encephalopathy and related diseases." *Emerging Infectious Diseases* 4 (1998): 390-94.

Postman, N. *Building a Bridge to the 18th Century.* New York: Vintage Books, 1999.

Quist, D. and I.H. Chapela. "Transgenic DNA introgressed into traditional maize landraces in Oaxaca, Mexico." *Nature* 414 (2001): 541-3. See also the debate about this paper published in *Nature* 416 (2002): 600-602; *Nature* 417 (2002): 897-8.

Robert, S. and U. Baumann. "Resistance to the herbicide glyphosate." *Nature* 395 (1998): 25-26.

Rural Advancement Foundation International. "Monsanto vs. Percy Schmeiser." *Geno-Types* (April 2, 2001).

Saul, J.R. *Voltaire's Bastards: The Dictatorship of Reason in the West.* New York: Vintage Books, 1992.

Schwartz, E. *Overskill.* New York: Ballantine Books, 1971.

Shulman, M.H. "More milk, fewer farmers." *The New Farm* (November/ December 1987): 28-29, 39-41.

Stolberg, S.G. "Two cheers for human cloning," *The New York Times*. December 2, 2001: WK4.

Teitel, M. and K.A. Wilson. *Genetically Engineered Food: Changing the Nature of Nature,* 2nd Edition. Rochester, VT: Park Street Press, 2001.

Tesauro, J. "Restoring wetland habitats with cows and other livestock." *Conservation Biology in Practice* 2 (2001): 26-30.

C. KRISTINA GUNSALUS

HUMAN SUBJECT PROTECTIONS

Some Thoughts on Costs and Benefits in the Humanistic Disciplines

I. INTRODUCTION: PROTECTION OF HUMAN SUBJECTS AND THE HUMANISTIC DISCIPLINES

The federal protection of human subjects of research in the United States has from the start been scandal-driven, shaped by media attention and political reactions. The first of many Congressional hearings were catalyzed by the revelations of the 1966 Beecher Report that described experiments on unsuspecting patients from the previous year of published medical literature,[1] the Tuskegee Syphilis Study that tracked the course of the disease untreated for 40 years in 400 African-American men even after treatment became available,[2] and advances in medical research that were challenging religious and societal concepts of life and death.[3] Regulations in this area have grown by fits and starts, with bouts of intense study and scrutiny following newsworthy events interspersed with quiescent periods. These cycles have had a self-perpetuating quality, for lack of follow-through on policy recommendations[4] and frequent under-funding of protection mechanisms during the fallow periods have heightened rhetorical levels and the sense of urgency in the active periods. As the result of the cyclic but inconsistent nature of this process, we have been left with regulations that both under-reach and over-reach, that are both too broadly and too narrowly applied. While some people are subject to "experimentation" without any federal protections because the activities in which they are involved are not federally funded, in other places, the reach of the regulations has been so far extended that innocent "interactions" posing no risk to

[1] Dr. Beecher published information on and citations to twenty-two experiments. Examples include Case 2, in which physicians withheld penicillin from servicemen with strep infections without their knowledge or consent (published in the *Journal of the American Medical Association*); Case 16 that involved the administration of live hepatitis virus to residents of a home for the retarded (including children) to study the etiology of the disease (*New England Journal of Medicine*); and Case 17 in which live cancer cells were injected into hospitalized elderly and senile patients to study their immunological responses (need publication information). Information summarized from Rothman.

[2] Jones, *Bad Blood*.

[3] Rothman, 148.

[4] See Appendix A, "Examples of Studies with Multiple Unimplemented Recommendations."

A. W. Galston and C. Z. Peppard (eds.), Expanding Horizons in Bioethics, 35-58.

anyone are regulated, scrutinized and restricted. Both are costly side-effects of a system out of balance.

We are now in the midst of another period of scrutiny catalyzed by scientific advances and ethical concerns about their applications—cloning, modifying human genetics and stem cell research—and sustained by reports of deaths in medical research such as that of Jesse Gelsinger at the University of Pennsylvania, and the consequent media coverage.[5] Once again, a "crisis management" dynamic[6] is in play, with the additional effect that some institutions have become hyper-cautious as the financial, legal and reputational costs of problems escalate. On the positive side, this attention can lead to needed improvements in human subject protections. But there is also danger that we will impose even more unneeded and costly regulatory burdens with little gain in protections. The equation is a delicate one and we seem to have gotten it wrong in some places.

Human Subject Protections in the Humanistic Disciplines

How did we come to be regulating "interactions" in the humanistic disciplines, when the original focus of the rules was on experimentation in biomedical and behavioral research? Such regulation in the humanistic disciplines seems to be the result of recent scandals in biomedical and behavioral research, which caused federal officials to focus intently on activities within their reach, namely, federally-funded research, which occurs primarily at universities. With each cycle of scandal-driven attention, universities, seeing the costs for mistakes at other places, have in turn ratcheted up the intensity of their internal scrutiny, to avoid program shut-downs and costly investigations and attention drawn by scandals. Both to be and to seem "ethical," universities have, over time, voluntarily extended the application of federal regulations from covering only research supported by federal funding to all research conducted at the institution or under its auspices. Thus have more and more university-based endeavors come to be subject to federal regulation.

The stakes are high because the "fit" of the regulations to the humanistic disciplines is not always good; the nature of scholarship in the humanities is such that thoughtless application of regulations appropriate to biomedical research can cause harm, not only to the scholarship, but also to important principles including First Amendment protections and academic freedom.

Examples of this conflict abound. An historian working on oral histories of the civil-rights movement was cautioned not to ask subjects about the laws they might have broken in the course of civil disobedience.[7] An Institutional Review Board (IRB, the term of art for human subject protection boards) administrator directed that an English professor's in-press article be withdrawn because it contained an account of a student who submitted a paper describing his alleged participation in a murder,

[5] National Bioethics Advisory Commission, "Ethical Issues in Human Stem-Cell Research" and National "Cloning Human Beings: Ethical Considerations"; "Researchers Seek Answers After Gene-Therapy Patient Dies"; and Elmer-Dewitt.

[6] Nishimi.

[7] "Protecting Human Beings: Institutional Review Boards and Social Science Research."

and the article had not been submitted for IRB review.[8] Joan Sieber and her colleagues have reported such "mis-regulation" as:

> ...a linguist seeking to study language development in a pre-literate tribe was instructed to have them read and sign a consent form...a political scientist purchased appropriate names for a survey of voting behavior (of people who had consented to such participation) and was initially required by their IRB to get the written informed consent from subjects before mailing them the survey; a Caucasian Ph.D. student, seeking to study career expectations in relation to ethnicity, was told by the IRB that African American Ph.D. students could not be interviewed because it might be traumatic for them to be interviewed by the student....[9]

And journalism faculty have found both their own and the investigative projects of their students limited or prohibited by IRBs concerned about embarrassment to prospective subjects.[10]

While we have several decades of close analysis starting from fundamental ethical principles to provide guidance to researchers and regulators in biomedical and behavioral research, there is no such body of work in the humanistic disciplines. Yet the federal research regulations have come to apply to "all" types of federally-supported research within regulated institutions. These regulations, while framed expansively, emerged from an analytical context limited to certain categories of research (namely, biomedicine), whilst excluding almost totally other forms of interactions with human subjects. The increasing application of the literal meaning of the regulations to endeavors where the effects have not been analyzed—i.e., the humanistic disciplines—is a matter needing immediate attention and close analysis.

The ethical principles underlying current regulations are so rooted in considerations related to invasive or risk-laden medical procedures that even discussions of behavioral research, which have been explicitly included within the regulatory scope since early in the consideration process, can seem "tacked" on. Again, this is rooted in a scandal-driven history: much of the consideration of behavioral research followed public debate about the Milgram authority/shocking experiments[11] and the Humphrey's "tearoom" research.[12] As a result, the initial

[8] Personal communication with the author. Details omitted to protect privacy.

[9] Sieber, Plattner, and Rubin, 1-4.

[10] "Protecting Human Beings: Institutional Review Boards and Social Science Research."

[11] In 1961-62 at Yale University, Professor Stanley Milgram recruited subjects to participate in research about "punishment and learning" through a newspaper ad. Subjects administered increasing levels of what they thought were electric shocks to the learners, who were actually part of the experimental team and did not receive shocks, although it appeared to the subjects that they did. In advance of the research, through consultation with psychiatrists nationwide, Milgram anticipated that a large majority of subjects would refuse to administer painful or harmful shocks to the "learners." In actuality, 65% of the subjects continued increasing the level of the shock to 450 volts in obedience to the researcher/authority figure, even when they could hear what they thought to be screams of pain from the "learners" and in the absence of any coercion other than the instructions of the researcher.

[12] For his 1970 Washington University Ph.D. thesis, Laud Humphrey conducted research on impersonal male sex in public restrooms in which some participants were interviewed at the time, and others only later. Those interviewed later were tracked through their license plate numbers, and Humphrey a year later (in disguise) contacted participants to conduct a "mental health" survey, resulting in findings about the demographics and motivations of participants. The research was hugely controversial at the

analyses of applicable ethical principles for regulation of human subjects were primarily focused on behavioral research in medical settings, with the remaining analyses driven by the scandals appended as the process unfolded.[13] The articulation of the ethical principles applicable to research with human subjects, and the very wording of the regulations themselves, trumpet their quantitative focus and orientation. The definition of "research" in the regulations, for example, hinges on endeavors designed to contribute to "generalizable knowledge," which is appropriate for biomedical research, but not necessarily scholarship in the humanities, where this can be read so broadly as to be meaningless.

Novel applications of existing regulations are also illuminating different aspects of their shortcomings, some of which are potentially more costly and perhaps less tolerable than previously thought. For example, procedural shortcomings in existing regulations—most notably the lack of any appeal mechanism from IRB decisions about permissible research activities—that are worrisome for any scholarly endeavor are even more so when applied, for example, to the investigations of journalism faculty and students, where they become a form of prior restraint particularly odious to our democratic systems. And disturbing examples are emerging of instances in which IRBs seem to have been used as tools by those opposed to the findings of some researchers.[14]

Before this cycle of attention closes, we have an opportunity to bring thoughtful consideration to how current regulations apply to scholarly activities in the humanistic disciplines and, perhaps, to improve the balance of protection and burden in the overall public policy equation. What is needed is to develop guidance on a range of questions, within and across disciplines and with the engagement of national disciplinary societies and other appropriate bodies such as the American Association of University Professors (AAUP) and Public Responsibility in Medicine and Research (PRIMR). The sections that follow provide a case study to illuminate the problem of oversight for the humanistic disciplines; an overview of existing regulatory scope and procedures; a summary of the policy issues presented by examples of over- and under-reaching of the current human subject protection regulations; and possible approaches for improving our protection system while taking these issues into account.

II. A DESCRIPTION OF THE PROBLEM: WRITING ABOUT TEACHING

Given their original limited focus to biomedical and behavior research, the definition employed in the federal regulations is that a "human subject" is a living person about whom "an investigator… conducting research obtains (1) data through intervention or interaction" with the individual, "or (2) identifiable private information." Clearly, the definition of "conducting research" here is key. We will return to that point

---

time, with a number of faculty members petitioning the President to withhold Humphrey's degree on the grounds that its deceptive elements were unethical.

[13] Levine.

[14] Tavris; Begley, B1.

later, because it has been largely skipped over as universities and IRBs have focused upon the "interaction" portion of the regulatory language. While the responsible federal oversight agency has recently concluded that oral history interviews are *not* covered by the federal regulations in a feat of bureaucratic logic (briefly, that they are not seeking to contribute to generalizable knowledge),[15] the assumption has been that all faculty members who are publishing are "conducting research" and so are covered by the regulation. Consider the application of the regulatory definition of "human subject" to the humanistic disciplines by the case of a faculty member writing about his or her professional activities.

When is teaching a class (or conducting research and then writing about the process) an "interaction" with a human subject? When a faculty member writes an essay about an experience teaching a class, when is this an autobiographical essay and when does it become research? Consider the regular essays in the *Chronicle for Higher Education* detailing the perspective of various academicians about different aspects of their own institutions, from hiring to departmental politics. If a faculty member writes about a particularly illuminating classroom exchange, or discourses anecdotally about the difference between freshmen now and twenty-five years ago, these are clearly "interactions" with human beings. Is it, or should it be, covered "research" such that an IRB can or should require advance review and approval of the project?

If a faculty member writes an essay about one semester's teaching experience at the end of a semester that contains anecdotes (suitably altered to protect individual identification), is that research? What if the faculty member writes several essays that are published throughout the semester? What if the arrangement to write the essays is made in advance of the semester and the faculty member constructs the syllabus to enhance interactions that might generate memorable anecdotes? What if the faculty member makes notes on each class session as an aid to the drafting of the essay? What if the professor records each class to save memorable phrases?

Serious professional and ethical issues can arise in writing about teaching and classroom interactions, but they are not necessarily issues illuminated by federal regulations about interacting with human subjects. Faculty members must grapple with how to treat deeply personal disclosures made by students,[16] stay within the boundaries of academic integrity, respect intellectual property, and adhere to the laws governing student privacy. These issues are receiving careful attention in the literature of some disciplines,[17] but there has been little consideration of the global considerations that may apply across disciplines. Thus, when we do arrive at the ethical issues implicated by the human subject regulations, humanistic disciplines seem to be lacking systematic consideration of the elements that move from personal reflection to "research" (e.g., a piece of writing by a faculty member about classroom experiences). By extension, we also lack clarity on how to weigh considerations of academic freedom and prior restraint.

[15] Office for Human Research Protection.

[16] Morgan.

[17] "Guidelines for the Ethical Treatment of Students and Student Writing in Composition Studies."

In thinking about how to define "research" in writing about teaching, the ends of the spectrum are relatively easy to identify: a faculty member who begins a semester with the intent of comparing student grades and outcomes in a skills course, for example, with overall grade point averages of each student, with the goal of correlating expertise in the skill with the remainder of the students' graded coursework, is surely conducting research. This would seem to be an "experiment" with "human subjects." The professor is approaching the class with the intent to collect data in a systematic way and to publish the results, presumably with the hope of being able to come to generalizable conclusions. Here some of the central issues are raised: there are ethical aspects of this conduct to explore, including the power relationship of the faculty member over the students, but given the nature of the potential harm, what levels of protection and regulatory action are appropriate? But between that faculty member's planned "research," and the other end of the spectrum where a faculty member is musing about a lifetime of teaching in an autobiographical fashion, the lines are fuzzy. In between those two extreme cases, authors "interact" with other humans; they must also approach their thinking and writing systematically (one presumes). This brings the activity under the current language of the federal regulations. What, exactly, are the characteristics that do or should make it "research" requiring IRB review? What moves an activity across the line? When do we care? And why?

If the driving purposes of our human subject protection system are to protect individuals from harm and to "be ethical," what are the harms from which we are protecting people in these settings? Are they on the same plane as protecting patients in clinical trials? Do they require the same levels of oversight and regulation?

Finally, in any number of forms of professional education outside medicine, clinical training is provided to students where both the student and the student's patient or client have the potential to become "subjects" when the supervising faculty member later writes about the educational process. Thus, in a clinic associated with a law school, when should the student and the client provide informed consent for the supervision provided by the faculty member?[18] Where students are being trained to become licensed social workers? Or doctors? Are the potential harms commensurate with the regulatory burdens we impose?

Clinical educators are increasingly alarmed with such IRB rulings that students may not, without informed consent, write about experiences with clients during internships, even for course papers to be submitted to and read only by supervising faculty members, and even with no identifying information.[19] And, because there are no appeal mechanisms for IRB decisions, there is no recourse when a local IRB adopts such an interpretation. The lack of consistency we are seeing in the application of regulations is likely rooted in the absence of well-understood and broadly accepted ethical principles and guidelines.

Until we can articulate the answers to these questions with clarity, we continue to run the risk that various IRBs will each invent their own answers for each activity

[18] Anderson, 1.
[19] Tarr.

presented for review, with little consistency. At the very least, this is an ineffective use of resources. At most, IRB conclusions (or those of their administrators, as many questions are not even presented to, or considered by, an IRB as a whole) that are inconsistent (or unreasonable or silly) will increase disrespect for compliance systems in universities, to the detriment of the important policy purposes that these systems serve.

III. A BRIEF SUMMARY OF HUMAN SUBJECT PROTECTION: HISTORY AND SCOPE

Federal regulation of research with human subjects in the United States is restricted to 1) research funded by certain federal agencies, 2) conducted at institutions filing assurances with the Office of Human Research Protection, and 3) studies to be submitted to the Food and Drug Administration for drug, biologics and device approvals.[20] These regulations look first, to the source of funding for the work, and second, to the site *where* the work is performed (rather than upon the risk/harm to the subjects regardless of funding or locale of performance). This last consideration is an issue to which we will return.

In the United States, most (but not all) federally funded research on humans is governed by a regulation now known as the Common Rule that has been in effect since 1974.[21] Although the National Institutes of Health issued the first *publicly available* federal regulations in 1966,[22] current regulations are rooted in the work of a congressionally mandated body, the National Commission for the Protection of Human Subjects of Biomedical and Behavioral Research. The National Commission was formed as a reaction to well-publicized scandal and controversy over scientific advances and it was strenuously resisted by the scientific community. Working from 1974 to 1978, the Commission produced a seminal report known as the "Belmont Report,"[23] identifying the ethical principles that still serve today as the foundation of all U.S. human subject regulation: "respect for persons," "beneficence", and "justice."[24] In practice, "the principle of *respect for persons* underlies the need to obtain informed consent; the principle of *beneficence* underlies

[20] Federal Food, Drug and Cosmetic Act.

[21] Code of Federal Regulations, Title 45.

[22] The work of the Advisory Committee on Human Radiation Experimentation (1994-95) reported that the Atomic Energy Commission and the Department of Energy issued regulations governing human experimentation in the 1940s and 1950s, but since the regulations were classified, few but very high-ranking officials ever knew of their existence. (See Moreno's essay in this volume for a discussion of human subjects protections in the military. *Ed.*)

[23] That report also cites the foundation provided by both the Nuremberg Code and the Helsinki Declaration.

[24] *Respect for persons* involves recognition of the personal dignity and autonomy of individuals, and special protection of those persons with diminished autonomy. *Beneficence* entails an obligation to protect persons from harm by maximizing anticipated benefits and minimizing possible risks of harm. *Justice* requires that the benefits and burdens of research be distributed fairly. National Commission for the Protection of Human Subjects of Biomedical and Behavioral Research, "The Belmont Report."

the need to engage in a risk/benefit analysis and to minimize risks to human subjects; and the principle of *justice* requires that subjects be fairly selected."[25]

The Common Rule grew directly from the Belmont report.[26] Of the twenty-three federal agencies that fund research, seventeen subscribe to the Common Rule, and the others are—and have been for some time—under executive order to come into compliance.[27] Why does this inconsistency across the federal government continue? Might the mismatch of regulation to disciplines outside the biomedical arena be a contributing factor? That is, perhaps one reason that it has been so hard to get all federal funding agencies "on the same page" is because those in endeavors outside the biomedical arena see the mismatch of the regulations to the nature of the work they support—and the general absence of ethical dilemmas in those areas that are helpfully addressed by current implementations of the Common Rule.

The Common Rule defines "research" very broadly, as "[a] systematic investigation, including research, development, testing and evaluation, designed to develop or contribute to generalizable knowledge."[28] In order to focus upon activities seen as holding the highest risk for the subjects of that research, it defines a number of categories of activities that are "exempt"[29] from review if certain procedures are followed, and a number of others that qualify for "expedited"[30] review on the theory that they present only minimal risk to the subjects. The concept and implications of assessing risk to subjects was the focus of much of the early ethical analysis of proposed regulatory systems. Yet, it is remarkable that neither "risk" nor "harm" are defined in the regulation.[31]

The Common Rule mandates that people who conduct research involving humans must establish local review and approval bodies known as Institutional Review Boards (IRBs). As a result, there are an estimated 3,000-5,000 IRBs in the United States.[32] Broadly speaking, wherever an IRB has jurisdiction, its review and approval are required before research involving human subjects can begin.[33] IRBs are intended to be qualified to assess proposed research both professionally and from a lay perspective, so that (theoretically, at least) the sensibilities of the community are consulted before research begins, not just in the aftermath of scandal. Thus, the Rule specifies that each IRB must have a

[25] National Institutes of Health.

[26] National Commission for the Protection of Human Subjects of Biomedical and Behavioral Research, "The Belmont Report."

[27] Gunsalus, "An Examination of Issues," D-6.

[28] Code of Federal Regulations, Title 45.

[29] Code of Federal Regulations, Title 45 § 46.101. See Appendix B.

[30] Code of Federal Regulations, Title 45 § 46.110. See Appendix B.

[31] Finkin.

[32] Estimates of the number of IRBs operating in the U.S. range from around 3,000 to more than 5,000. Puglisi.

[33] Some federal agencies have come to require evidence of IRB approval before a proposal will be considered for funding, a development leading to many complaints, as it increases the workload of IRBs in that they must review research that may never be performed because it does not get funded.

> diversity of the members, including consideration of race, gender, and cultural backgrounds and sensitivity to such issues as community attitudes, to promote respect for its advice and counsel; and... shall include at least one member whose primary concerns are in scientific areas and at least one member whose primary concerns are in nonscientific areas....Each IRB shall include at least one member who is not otherwise affiliated with the institution...[34]

By virtue of the bureaucratic mechanism by which universities commit to comply with federal requirements and as a result of pressure from the scandal cyles over the years, most agree to extend the federally mandated rules to *all research* conducted at their institutions. This encompasses research conducted by students, faculty, and any employees under university auspices. This broad assent to regulatory jurisdiction brings virtually all work conducted at universities under federal regulation, regardless of its source of funding.

Since virtually all consideration and enforcement of human subject policy has been focused upon biomedical and behavioral research, it may seem simple to rectify the over-application of human subject regulations in the humanistic disciplines. Yet simplicity eludes us. From the onset, it has been clear since the Common Rule was intended to apply to "all" research. It was just that other areas of research were not generally federally funded at the time and the applications were not as carefully analyzed as was work in the biomedical arena. Indeed, the National (Belmont) Commission and its successor the Presidential Commission (1980-1983), and the Advisory Committee on Human Radiation Experimentation (1994-1995), were all chartered to examine "biomedical and behavioral research" and focused their work in those areas. The National Bioethics Advisory Commission (1995-2001) was similarly charged with advising and making recommendations on "bioethical issues arising from research on human biology and behavior." Until the work of the National Human Research Protection Advisory Committee (NHRPAC) in 2001, I cannot find any federal commission or agency that examined the application of federal regulations to humanistic disciplines. The emergence of regulation of humanistic research is recent.

In 1977, the Presidential National Bioethics Advisory Commission (NBAC) maintained that "No person in the United States should be enrolled in research without the twin protections of informed consent by an authorized person and independent review of the risks and benefits of the research."[35] Whether the researcher is a neurologist or a poet, if the research involves interaction with a human subject, this suggests that work should be subject to regulation. Legislation to this effect is introduced regularly on Capitol Hill. The concerns are not idle and the significance of the shift from the Belmont Report's focus on "subjects of biomedical and behavioral research" to the current focus on "research" writ large should not be underestimated. This expansion from biomedical and behavioral research to "all" research raises questions of what our public policy goals are and should be with regard to human subjects protection.

[34] Code of Federal Regulations, Title 45 46.107.

[35] National Bioethics Advisory Commission, Full Commission Meeting.

David Rothman's *Strangers at the Bedside: A History of How Law and Bioethics Have Changed Medical Decision-Making* provides a valuable foundation for considering policies governing human subjects of research. Rothman primarily focuses on the trends that coalesced to wrest medical decision-making control from physicians (especially at the beginning and end of life), and toward patients, families, lawyers and courts. Nonetheless, his work illuminates the forces that have shaped most U.S. research regulation in recent decades, including a steadily growing mistrust of experts. The prevailing cynicism about "experts" will immensely complicate the resolution of current questions. In particular, this developing cynicism suggests that referring development of standards or responsibility for oversight solely to disciplinary communities for internal deliberations will not be an acceptable outcome. Instead, consensus will need to be developed systematically and endorsed across disciplines and by national advisory groups that include members from outside the affected fields. The challenge is to assure that all participants in the process have a sufficient base of understanding, such that the outcome is sound public policy.

In a 1998 paper commissioned by the National Bioethics Advisory Commission (NBAC), I addressed some of the dilemmas we face in balancing regulatory costs and benefits in the area of human subjects of research. Due to threats of litigation and society's mistrust of experts, the incentive is toward ever-increasing scrutiny and regulation of research involving human subjects. This often occurs without examination of the correlative costs, either in direct regulatory infrastructure or in scholarship forgone—especially if one attempts to add in the question of "how will these changes affect research outside the biomedical arena?" Usually, that question is not even considered, as the focus continues to be upon biomedical work, again driven by scandals. In light of the increasing problems seen from the imposition of the biomedical model on the humanistic disciplines, aggregated with the other issues raised in recent years, it seems past time to rebalance the regulatory equation.

To do so, we must 1) identify with specificity the harms/risks against which we seek to protect subjects; and 2) articulate the ethical, principled basis for governmental and/or institutional provision of those protections. In pursuing those analyses, we must be able to define which activities constitute "research" subject to regulation, who "human subjects" are, and why we are choosing to devote resources to those ends. But first, there are some definitional problems to address.

IV. THE COMPLICATION OF DEFINITIONS

In order to articulate the problems and risks of research, we need to know what "research" is and who the "human subjects" are for whom we need to prevent exposure to risks and harms. Recall that "research" is "a systematic investigation, including research development, testing and evaluation, designed to develop or contribute to generalizable knowledge." This definition grew out of the Belmont Report and has been in use for more than thirty years. The application of regulation is always (or should be) contingent upon whether an activity constitutes "research" under the federal definition. Sensibly enough, the federal rules provide that wherever a question exists on this point, it must be resolved by an independent individual (not the individual pursing the activity) to protect against conflicts of interest in the determination.[36] The process by which these decisions are reached is left to the institution's IRB. In practice, there is generally no appeal from the determinations of an IRB or its administrator. We will revisit this last point later.

Once it has been determined that an activity *is* research, the rest of the definition then applies: a human subject is "a living individual about whom an investigator (whether professional or student) conducting research obtains (1) data through intervention or interaction with the individual, or (2) identifiable private information."[37]

The regulations do *exempt* specified activities, such as comparisons of the efficacy of educational techniques, observations in public settings, and examination of existing records where no individual will be specifically identified. (The validity of the premises upon which many forms of research in educational settings were exempt have been called into question, and may warrant re-examination.)[38] Other categories of research qualify for *expedited* review on the grounds that they pose no more than minimal risk to the subjects of research. The regulations also specify how IRBs should be constituted and how they should operate: how they keep their records; establish criteria and procedures for review; characterize elements of informed consent; and address a number of more specialized topics, especially those involving vulnerable populations (prisoners, fetuses, children).[39]

It is generally left to the IRB to decide whether or not something is "research." One major rule of thumb regularly used by IRB members to determine whether an activity is "research" is whether the efforts were undertaken with an eventual eye to publication; this mode of determination stems from the ethical analyses commissioned for the Belmont Report.[40] This approach works fairly well in biomedical research: if a practitioner or scientist planned a project, began it and assembled data in a way that would support publication in the peer-reviewed literature, the likelihood is reasonably high that some kind of "research" was underway. Consequently, those upon whom hypotheses were tested had an ethical right to knowledgeable participation. This approach can be problematic, however, when applied in other settings.

One of These Things is Not Like the Others: Problems Specific to Research in the Humanistic Disciplines

It is hard to argue that any writing by a humanist or scholar in the humanistic disciplines is not, broadly speaking at least, intended to contribute to "generalizable knowledge." When an English professor writes about interactions with students and

[36] Code of Federal Regulations, Title 45 § 46.102.

[37] Ellis.

[38] Howe and Dougherty.

[39] Code of Federal Regulations, Title 45 § 46.101. See Appendix B.

[40] Levine.

muses about how that experience shaped his thinking about teaching, does that not contribute to generalizable knowledge? What about a journalist's exposé of the excesses of local government?

There are multiple examples of this dilemma: is it, for instance, research when physicians write newspaper columns about their experiences with and reactions to various patients or diseases? Is it research when a journalist interviews a series of public officials (clearly these are "interactions" with "individuals") and publishes a story about corruption at city hall? If not, why would the same work pursued by a graduate student in a journalism program become research? And why would we consider it "research" when a faculty member writes in a professional journal about classroom teaching experiences? The logical extension of the "intended for publication" guideline is that all human activities that end in (or are intended for) publication are subject to federal oversight in the same way as biomedical endeavors are. Indeed, the latter two examples have both been defined as "research" in universities by institutional officials. Is this the correct application and desired outcome? Is it one we should support? Does this approach contribute to the public policy goal of protection of human subjects?

One troubling aspect of addressing these questions is that the determination often correlates to the location of the research activity—not just the source of funding. If a pollster calls and asks about your preferences for laundry detergent or political candidates, that is not "research" governed by the federal regulations—unless the pollster happens to be a university student or faculty member. Then, the exact same activities *do* constitute "research" on "human subjects" and must receive IRB approval (or exemption confirmation) before they begin—otherwise, the data collected may not ethically be published.[41] A clear problem emerges: Should activities that are not research if performed in the private sector become research for the sole reason that they are performed in a university setting? While there may be other aspects of their performance in universities that might transform them into research, the setting alone surely does not. The obverse situation also raises questions. In my NBAC paper, I suggested that a published comparison of two different techniques for facial surgery undertaken by a cosmetic surgeon was research on human subjects, and that the "patients" deserved protection, regardless of the fact that the research was privately funded, was conducted outside a university, and thus was not presently regulated as "research."

V. REGULATORY AND PROCEDURAL ISSUES

Inconsistent Regulation

For my NBAC paper, I assembled examples of both unregulated research carrying serious risks to its participants as well as examples of activities that have been placed under current or proposed regulatory language but that do not expose

[41] Gunsalus, "An Examination of Issues."

participants to risks that would seem to warrant oversight.[42] That is, there are activities fitting the definition of "research" involving human subjects—and about which complaints have been filed with federal agencies—that are not covered by federal protective regulations, regardless of the "risk" or "harm" to which they expose subjects. And on the other hand, there are activities subject to regulation that seem to have little harm or risk for the participants. The examples I assembled were based primarily upon Freedom of Information Act Requests filed with the Office of Protection from Research Risks (OPRR)[43] involving complaints over which the office had no jurisdiction.

Examples of the former category (presently unregulated but risky activities) include genetic tests; *in vitro* fertilization research; and experimental surgical procedures developed without IRB oversight (because the work is performed at unregulated for-profit or not-for-profit entities); research funded by a pharmaceutical company involving children of short stature identified by private physicians for payment; and "fright response" research in which subjects were subjected to disturbing stimuli.[44]

At the same time, activities that seem low-risk are regulated, primarily because they are performed at universities or by faculty or students. These include any number of interviewing protocols conducted under disciplinary ethical protocols (e.g., anthropology, journalism) or aspects of clinical education and scholarship that involve writing about teaching and experiences in the classroom, as discussed above. It is crucial to note that none of these "regulated" activities have engendered scandal or controversy and, absent specifically unethical aspects (deception, fraudulent conduct, etc.), seem unlikely to do so. Of course, an argument could be made that these activities are ethical because they have been vetted through an IRB. Whether or not that objection holds in all cases (I believe it does not), it is clear that we do not clearly define the risks and harms against which we wish to protect subjects of research—especially outside the biomedical realm. As a result, our regulatory efforts both over- and under-reach. Surely a sounder public policy would work to cover the highest risk activities before sweeping into the scope of regulation activities that are low risk. Additional analysis of regulation of humanistic research, on a level at least as rigorous as that devoted over decades to biomedical and behavioral research, seems warranted. This would assure that both the costs and benefits of regulations serve intended public policy ends.

Procedural Issues

In addition to the substantive challenges presented by regulation of research in the humanistic disciplines, there is increasing controversy across the country on procedural aspects of IRB oversight of research. Because of the lack of consistency in IRB function and decision-making, researchers have contended that IRBs have

[42] Ibid., 24-28.

[43] This is the precursor office to the current Office of Human Research Protections within the Department of Health and Human Services.

[44] Gunsalus, "An Examination of Issues," D14-15.

been used as tools to attack work with unpopular conclusions; that IRB decisions are not subject to any form of appeal or review; that inter-university research proposals must be approved by multiple IRBs that do not communicate with each other and that may come to different conclusions; and that the traditional mechanism for determining whether an activity is exempt from IRB review is insufficient (since it requires submitting the proposal to the IRB, or its administrator, without an appeal mechanism for these *ad hoc* decisions). Decisions made by IRBs without an adequate or clear reasoning mechanism are among those creating controversy and distrust among researchers in the humanities.

Carol Tavris, a fellow of the American Psychological Association, suggests (in an article describing the tribulations of two psychologists studying recovered memory) that:

> Today, many of the IRBs originally established to protect subjects have instituted so many Byzantine restrictions and rules that even good scientists cannot do their work. Some have become fiefdoms of power—free to make decisions based on caprice, personal vendettas, or self-interest, and free to strangle research that might prove too provocative, controversial or politically sensitive.[45]

VI. HUMANISTIC DISCIPLINES: DISCIPLINARY CONCERNS AND PROPOSALS

Recall the puzzling questions raised in the discussion of writing about teaching. When such questions emerge in the humanistic disciplines, it seems useful to concentrate first on trying to articulate what activities comprise "research" and to codify guidance on that point to improve consistency of IRB judgments. It is possible that many of the current controversial rulings by IRBs that raise matters of principle could be resolved by disseminating guidance on, first, the substance and criteria for decision-making and, second, improved and consistent procedures for reaching such decisions. Failing that, better guidelines for how the federal regulations and local IRBs relate to scholarship in the humanities, journalism and some branches of social science would solve many of our present quandaries. There is one overarching question to ask of the goals of the protection system: who are we aiming to protect, and from what?

The American Association of University Professors (AAUP), in a 2001 report on the oversight of social science research by IRBs, highlighted a number of these issues.[46] Responses to that report, including comments made by panelists during the January 2002 NHRPAC meeting and published in *Academe*, continue the discussion. These inquiries explore the emerging view that "the government's regulations, known as the Common Rule, as applied by campus institutional review boards to humanities and social science research, sweep too broadly."[47] The AAUP initiative explores the problems that have arisen from the application of the biomedical

[45] Tavris.
[46] American Association of University Professors.
[47] "Should All Disciplines Be Subject to the Common Rule?"

research paradigm to ethnographic research, to oral history interviewing, and to the teaching of journalism and mass communication. The participants propose a variety of solutions, from specialized IRBs with appropriate expertise for reviewing work in these disciplines, to discipline-based guidelines for best practices and expedited reviewing procedures,[48] to excluding all research from oversight unless it poses "a risk of physical harm."[49] More recently, a Canadian national body has proposed that their equivalent of IRBs should be required to demonstrate "identifiable harm" before regulations apply.[50]

Some of these proposed solutions are more realistic than others. Discipline-based guidelines are sorely needed and should provide local IRBs with guidance on ethical and practical matters in areas that may frequently be beyond the expertise of their members. At larger institutions, specialized IRBs may be both practical and possible, although we should carefully consider the possible advantages and disadvantages of this approach. In addition to concerns about inconsistent application, one question is what might be lost by reducing the cross-fertilization across disciplines in the discussion of protection of human subjects. The proposal to exclude all research that does not pose a risk of physical harm seems unlikely and naïve in light of the history in this area. It may be, though, that a shift to the concept of "identifiable harm" holds promise and should be explored.

The AAUP conversations, moreover, may be conflating issues that should be examined separately. First, we must be able to articulate the point at which work performed by university-based teachers and scholars becomes "research" warranting review and oversight by an IRB. This then invokes the second question: Who are the human subjects, and what are the risks and harms to which they are susceptible? Only then should the third question arise, as to what provisions or guidelines should apply to that research.

There are further disciplinary complexities, as well. Whenever the conclusion is reached that an activity is "research," by what guidelines should it be determined that the activity falls under the exemption for educational research? These issues have yet to be considered in any comprehensive way.

Current exemptions encompass: "(1) Research conducted in established or commonly accepted educational settings, involving normal educational practices, such as (i) research on regular and special education instructional strategies, or (ii) research on the effectiveness of or the comparison among instructional techniques, curricula, or classroom management methods."[51] Within the literature on educational research, some have questioned whether the premises underlying the exemption for pedagogical research are valid, or were when they were drafted. Howe and Dougherty, for example, note that the special exemptions for educational research were formed before qualitative research methods were introduced, and suggest that aims and methods are a more important way to distinguish types of research than topics or settings. It should be noted that the current official IRB

[48] Ibid.

[49] Ibid.

[50] "Giving Voice to the Spectrum," 6.

[51] Code of Federal Regulations, Title 45 § 46.101, et seq.

Guidebook issued by the federal agency with enforcement responsibility has only one paragraph on "fieldwork" and one paragraph on "social policy experimentation," in all of its eleven (total) paragraphs on behavioral research, out of hundreds of pages.[52] This scant guidance focuses on the intimacy and open-endedness of qualitative research as requiring special scrutiny and perhaps new regulatory approaches.

The circumstances that gave rise to the educational exemptions also included the assumption that educational institutions (especially elementary and secondary schools where the students are generally under 18 years old) already exercised oversight over research performed in their settings, and that additional review by IRBs could be redundant, an operating assumption about which questions can be raised. Other rationales for the exemption focused on the very low risk nature of much educational research, and also strayed into the research vs. therapeutic treatment analysis used by the National Commission in its Belmont Report and commissioned papers. Perhaps these assumptions also warrant re-examination.

Another issue bearing consideration is the frequency with which anonymity and confidentiality are confused in the literature on educational research. Anonymity, privacy and confidentiality are each separate principles that address different issues. To use them interchangeably, as appears to be occurring in some debates on the ethics of writing about teaching—as has occurred in some of the internal discussions within the Modern Language Association, for example—complicates coming to a crisp and clear articulation of the ethical challenges to be addressed and the array of possible approaches to them.[53]

Possible Solutions?

In 1998, I came to the conclusion, perhaps too quickly, that a fundamental change in the federal definition of "research" was not warranted. The practical difficulties alone in changing the definition upon which our entire regulatory system rests are immense. Achieving consensus for a change in such a fundamental, longstanding definition, which involves researchers from all disciplines, is a daunting task. And yet the problems with oversight in the humanities make clear that the *status quo* is insufficient and that these unique concerns must be addressed.

One way to circumvent the problem of attempting to redefine "research" might be to discuss what covered research is *not.* Things that are *not* covered research would not be subject to oversight and regulation in the first place. If we embark on such a *definition by exclusion*, it would be important to consider the ramifications and possible unintended consequences of labeling certain activities as scholarship without *federal* regulation. (Institutional and disciplinary regulation are always still possible.) If we are uncomfortable "defining out" certain categories of research, perhaps we may suggest that development of *additional exemptions* would be a stronger approach than to "define out" certain activities in totality. That is, if we

[52] Howe and Dougherty.
[53] National Institutes of Health.

concede that some activities are "research," but that we do not believe they pose sufficient risks to adults, might we say without exception or prior review that they warrant an *ex-ante* protection system?

Possible categories of research needing careful analysis in this respect include ethnographic studies, oral history,[54] survey research (on the grounds that participants consent by filling out the surveys) and journalism. These are all activities that involve interviews or "interactions" with individuals for which decisonally-capable adults ought to be able to understand and assess—and accept—the "risks" of participation (whatever they might be) without paternalistic protection from an IRB.[55] This presumes that the activities are ethically and correctly conducted according to professional disciplinary standards. Others may claim that disciplinary standards have no regulatory clout, so IRBs are necessary watchdogs. However, institutions certainly have tools for responding to unprofessional and unethical behavior on the part of their members, and perhaps for abuses of those standards, the recourse ought to be made available through an *ex-post* complaint mechanism, but not an *ex-ante* review.

Along with the question of exemption is the issue of expedited review. If we believe an activity is "research" that should *not* be exempt from IRB review, who decides—and how—whether it qualifies for "expedited" review? In this form of review, only the IRB chair (or designated member) and administrator review the research protocol, generally on the grounds that the research poses minimal risk to the participants. Is it possible to assemble guidelines that clarify categories of research to which only expedited review requirements ought apply? Can we refine procedures so these responsibilities can be more broadly distributed within institutions? Can the ethical principles implicit in the Common Rule be so interpreted and applied? If so, it seems important to systematize, on a national basis, the concepts and methods so we can avoid the current system of *ad hoc* application within each individual IRB. Consistent definitions and review criteria grounded in disciplinary consensus that are followed by all IRBs would alleviate considerably some of the more serious problems now seen.

Several disciplines are working on guidelines internal to their discipline, as the AAUP case illustrates, but this does not go far enough. We need a multidisciplinary assessment of guidelines. Developing ethical principles that apply to educational researchers, for example, may or may not translate well to social psychologists studying organizational behavior in corporate workplaces or to the issues that arise in ethnographic studies in international settings. Only by focusing on the ethical issues that arise in methodological groupings, with multidisciplinary collaboration, will we be able to devise generalizable rules that can guide IRBs across the country. Otherwise, we will be left with either a highly particularized, university-specific regulatory system (the current situation) or a set of sweeping "one size fits all" federal statements that fail to comprehend the complexity of these issues.

As one step in this direction, an interdisciplinary working group at the University of Illinois has tentatively developed an overarching set of categories within which

[54] Since exempted by the Office of Human Research Protections, September 22, 2003.

[55] Gunsalus, "Thinking About Two People Talking."

these questions might be asked.[56] Those categories are: a) survey research; b) observational/interactional research (e.g., ethnography); c) biographical work, including oral history and reality-based fiction); d) autobiographical work (including writing about teaching); and e) interventional research methodologies, including those that employ non-invasive as well as invasive protocols.

For each methodology examined, there are three core questions to be addressed in formulating recommendations about which research ought to be subject to what level of IRB advance review and approval. 1) In work using this method, what constitutes research and experimentation under the federal definition? 2) What turns a person into a "human subject" in this area? We all interact with other people every day. How do we distinguish between interactions that do and do not require regulation? Is the "intent to publish" rule of thumb a fair or prudent standard? And 3) From what "risks" or "harms" are we protecting potential subjects, and why? And, of course, there is the underlying question of whether work conducted in university settings should be subject to regulations that are not imposed on the rest of society; or conversely, if the rest of society should be subject to the same ethical controls as those within the IRB system.

Until we can develop broad consensus across scholarly disciplines and produce guidance documents for IRBs, we continue to run the risk of wasteful use of valuable time and expertise for little demonstrable gain, to increase disrespect for important protective systems, and to delay or even impede altogether the pursuit of important research. Even the *Wall Street Journal* has commented that

> IRBs…are cracking down on social sciences, where the risk to volunteers amounts to hardly more than bruised feelings…Surely we can protect people in medical trials without strangling legitimate social-science research.[57]

Procedurally, we must address the composition of IRBs as well as the procedures by which determinations are made and reviewed—or not. Do we need specialized IRBs that focus on limited research areas (i.e., specific methodologies or fields) if we develop better guidance on what constitutes research and guidelines developed with disciplinary input that helps to address some of the most vexing issues and that provides first ethical principles for assessment of less-commonly arising problems? Is more oversight apparatus really a good use of resources? Does the "problem" justify this?

One approach might be to develop models for different sizes and types of institutions. What activities constitute "research" could be encompassed in this

[56] The Illinois project involves a set of structured, multi-disciplinary conversations designed to address and define the lines between research and scholarship in the humanistic disciplines, much the way that medicine has worked to define the difference between treatment and research. For example, what scholarly activities constitute "research" on human subjects that should be subject to advance review and approval? When an activity falls within the regulatory scope, what should the review process be and who should be involved? What disciplinary guidelines exist or need to be developed to guide these developments? At a national, invitational conference held in April 2003, these structured conversations were held across a variety of disciplines. The Illinois White Paper is forthcoming on these and related issues to continue the dialogue within and across disciplines, all of which are designed to affect how research outside the biomedical arena is reviewed and overseen.

[57] Begley.

model, since different sizes and types of institutions will have different types of research. Factors to consider include: How should we preserve the principle that decisions about what counts as "research" should not be made by the individual whose work is under consideration? And, on the opposite extreme, how do we place a system of checks and balances on the IRB or its administrator – perhaps by implementing a system of recourse or appeal? The new accrediting body for human subject protection programs, the Association for the Accreditation of Human Research Protection Programs, may well provide information and possibilities in this area as its accreditation criteria evolve.

It seems uncontroversial to assert that low-risk activities performed at universities or other covered entities should be a lower priority for regulators than high-risk activities outside the reach of present regulations. Yet with present trends for campus-based IRBs to encompass more activities under their review and with raised standards for review, we seem to be heading in the wrong direction. Various legislative proposals to expand protections for human subjects similarly over-reach, and leave low-risk activities over-regulated.[58]

VII. CONCLUSIONS

IRBs across the country need clear, consistent guidance about whom we are seeking to protect against what risks, why, and how to accomplish that goal. IRBs should not spend their resources reviewing activities that pose no more risk than is commonly encountered in daily life, and we should provide examples and standards so that more consistent, predictable decisions can be made nationwide. To this end, we must continue discussions within and across fields and disciplinary communities to develop common understanding of the goals, purpose and implementation of our necessary regulations. These conversations should involve disciplinary societies, national advisory committees and individual scholars. Recommendations must be both feasible and rooted in ethical principles.

Once (if) these conversations result in a reasonable national consensus, we will need both a mechanism for recording the outcome and for educating local IRBs. In this process, we must consider whether there are some areas that should not be subject to pre-approval procedures, but instead should have post-research complaint mechanisms as the appropriate check and balance for problems that arise.

We need an analytical framework that improves the current cost-benefit ratio of our protection systems for human subjects of research, especially regarding scholarship outside the biomedical realm. This framework must be applied using sound procedures that are seen as fair, even-handed and content-neutral. Without such structural revisions, we will continue to have a system that that imposes costs disproportionate to benefits in a large number of scholarly areas, and if not corrected, that will call our entire regulatory apparatus into disrepute.

[58] Research Revitalization Act of 2002.

Acknowledgements: Many thanks to Christiana Peppard for her patience and her helpful editing throughout a long development process. Many others commented helpfully, and to them I also owe thanks.

REFERENCES

American Association of University Professors. “Protecting Human Beings: Institutional Review Boards and Social Science Research.” *Academe* (May/June 2001).

Anderson, Paul. "Simple Gifts: Ethical Issues In The Conduct of Person-Based Composition Research." *College Composition and Communication* 49, no. 1 (February 1998).

Beecher, H.K. “Ethics and Clinical Research,” *New England Journal of Medicine* 274 (1966): 1354-1360.

Begley, S. “Science Journal.” *The Wall Street Journal* (November 1, 2002): B1.

Code of Federal Regulations, Title 45. Public Welfare Department of Health and Human Services, National Institutes of Health, Office For Protection from Research Risks, Part 46: “Protection of Human Subjects.” Revised November 13, 2001; Effective December 13, 2001.

Ellis, G.B. “Exempt Research and Research that may Undergo Expedited Review.” Office for Protection from Research Risks (May 5, 1995).

Elmer-Dewitt, P. “Cloning: Where Do We Draw the Line?” *Time Magazine* (November 8, 1993).

Federal Food, Drug and Cosmetic Act. 21 USCA §301 et seq. (1938).

Finkin, Mathew W. “Academic Freedom and the Prevention of Harm in Research in the Social Sciences and Humanities.” *Illinois Conference Proceedings* (2003). Available on conference website: http://www.law.uiuc.edu/conferences/humansubject/index.asp

“Guidelines for the Ethical Treatment of Students and Student Writing in Composition Studies.” *Conference on College Composition and Communication* 52 (2000): 485-490.

Gunsalus, C.K. “An Examination of Issues Presented by Proposals to Unify and Expand Federal Oversight of Human Subject Research.” National Bioethics Advisory Commission paper. September 1998. D-6.

________. “Thinking About Two People Talking.” *Ethics and Behavior* (forthcoming).

Howe, K.R. and K.C. Dougherty. “Ethics, Institutional Review Boards, and the Changing Face of Educational Research.” *Educational Researcher* 22, no. 9 (1993): 16-21.

Humphreys, Laud. *Tearoom Trade*. Chicago: Aldine, 1970.

Jones, J.H. *Bad Blood: The Tuskegee Syphilis Experiment.* New York: New York Free Press, 1993.

Levine, R.J. "The Boundaries Between Biomedical or Behavioral Research and the Accepted and Routine Practice of Medicine." *The Belmont Report: Ethical Principles and Guidelines for the Protection of Human Subjects of Research.* Washington, D.C.: U.S. Government Printing Office, 1978.

Morgan, D. "Ethical Issues Raised by Students' Personal Writing." *College English* 60 (March 1998): 318-325.

National Bioethics Advisory Commission. "Cloning Human Beings: Ethical Considerations." June 1997.

________. "Ethical Issues in Human Stem-Cell Research." September 1999.

________. Full Commission Meeting. Arlington, Virginia: May 17, 1997.

National Commission for the Protection of Human Subjects of Biomedical and Behavioral Research. "The Belmont Report: Ethical Principles and Guidelines for the Protection of Human Subjects of Research." DHEW Publication No. (OS) 78-0008.

National Institutes of Health, Office of Extramural Research, Office for Protection from Research Risks. *Protecting Human Research Subjects: Institutional Review Board Guidebook.* 1993.

Nishimi, Robyn Y. (Senior Associate, Office of Technology Assessment). Testimony at the Hearing Before the Legislation and National Security Subcommittee of the Committee on Government Operations, U.S. House of Representatives. Washington, D.C.: U.S. Government Printing Office, September 28, 1994.

Office for Human Research Protection. Guidance Letter. September 22, 2003.

"Protecting Human Beings: Institutional Review Boards and Social Science Research." *Academe* (May/June 2001).

Puglisi, Tom (Office for Protection from Research Risks). Personal communication. September 1998.

"Researchers Seek Answers After Gene-Therapy Patient Dies." *The Pennsylvania Gazette* (November/December 1999).

Research Revitalization Act of 2002. (§ 3060).

Rothman, D.J. *Strangers at the Bedside: A History of How Law and Bioethics Transformed Medical Decision Making.* New York: Basic Books, 1991.

"Should All Disciplines Be Subject to the Common Rule?: Human Subjects of Social Science Research." *Academe* (May/June 2002): 62-69.

Social Sciences and Humanities Research Ethics, Special Working Committee to the Interagency Advisory Panel on Research Ethics. "Giving Voice to the Spectrum." June 2004.

Sieber, J., S. Plattner, and P. Rubin. "How (Not) to Regulate Social and Behavioral Research." *Professional Ethics Report* 2 (Spring 2002): 1-4.

Tavris, C. *Skeptical Inquirer* 26 (July/Aug 2002).

Tarr, N.W. "Clients' and Students' Stories: Avoiding Exploitation and Complying with the Law to Produce Scholarship with Integrity." *Clinical Law Review* 5 (1998).

APPENDIX A:

Examples of Studies with Multiple Unimplemented Recommendations

Office of the Inspector General, Department of Health and Human Services. "Institutional Review Boards: Their Role in Reviewing Approved Research" (OEI-01-97-00190); "Institutional Review Boards: Promising Approaches" (OEI-01-98-00191); "Institutional Review Boards: The Emergence of Independent Boards" (OEI-01-07-00192); "Institutional Review Boards: A Time for Reform" (OEI-01-97-00193). Washington, D.C.: U.S. Government Printing Office, June 1998.

"Report and Recommendations of the National Commission for the Protection of Human Subjects of Biomedical and Behavioral Research," *The Belmont Report: Ethical Principles and Guidelines for the Protection of Human Subjects of Research.* Washington, D.C.: U.S. Government Printing Office, 1978.

"Report of the President's Commission for the Study of Ethical Problems in Medicine and Biomedical and Behavioral Research," *Protecting Human Subjects*. Washington, D.C.: U.S. Government Printing Office, 1981.

"Report of the General Accounting Office to the Ranking Minority Member, Committee on Governmental Affairs, United States Senate," *Scientific Research: Continued Vigilance Critical to Protecting Human Subjects.* Washington, D.C.: U.S. Government Printing Office, 1981.

United States Government, Human Radiation Interagency Working Group. "Building Public Trust," *Actions to Respond to the Report of the Advisory Committee on Human Radiation Experiments.* Washington, D.C.: U.S. Government Printing Office, October 1995.

APPENDIX B:

Basic Policy for the Protection of Human Subjects. Department of Health and Human Services, 45 CRF §46.101 and §46.110. Revised June 18, 1991; Effective August 19, 1991.

Subpart A: Federal Policy for the Protection of Human Subjects.
"b) Unless otherwise required by Department or Agency heads, research activities in which the only involvement of human subjects will be in one or more of the following categories are exempt from this policy:

(1) Research conducted in established or commonly accepted educational settings, involving normal educational practices, such as (i) research on regular and special education instructional strategies, or (ii) research on the effectiveness of or the comparison among instructional techniques, curricula, or classroom management methods.

(2) Research involving the use of educational tests (cognitive, diagnostic, aptitude, achievement), survey procedures, interview procedures or observation of public behavior, unless: (i) information obtained is recorded in such a manner that human subjects can be identified, directly or through identifiers linked to the subjects; and (ii) any disclosure of the human subjects' responses outside the research could reasonably place the subjects at risk of criminal or civil liability or be damaging to the subjects' financial standing, employability, or reputation.

(3) Research involving the use of educational tests (cognitive, diagnostic, aptitude, achievement), survey procedures, interview procedures, or observation of public behavior that is not exempt under paragraph (b)(2) of this section, if: (i) the human subjects are elected or appointed public officials or candidates for public office; or (ii) Federal statute(s) require(s) without exception that the confidentiality of the personally identifiable information will be maintained throughout the research and thereafter.

(4) Research involving the collection or study of existing data, documents, records, pathological specimens, or diagnostic specimens, if these sources are publicly available or if the information is recorded by the investigator in such a manner that subjects cannot be identified, directly or through identifiers linked to the subjects.

(5) Research and demonstration projects which are conducted by or subject to the approval of Department or Agency heads, and which are designed to study, evaluate, or otherwise examine: (i) Public benefit or service programs; (ii) procedures for obtaining benefits or services under those programs; (iii) possible changes in or alternatives to those programs or procedures; or (iv) possible changes in methods or levels of payment for benefits or services under those programs.

(6) Taste and food quality evaluation and consumer acceptance studies, (i) if wholesome foods without additives are consumed or (ii) if a food is consumed that contains a food ingredient at or below the level and for a use found to be safe, or agricultural chemical or environmental contaminant at or below the level found to be safe, by the Food and Drug Administration or approved by the Environmental Protection Agency or the Food Safety and Inspection Service of the U.S. Department of Agriculture."

§46.110
Expedited review procedures for certain kinds of research involving no more than minimal risk, and for minor changes in approved research:
"(a) The Secretary, HHS, has established, and published as a Notice in the Federal Register, a list of categories of research that may be reviewed by the IRB through an expedited review procedure. The list will be amended, as appropriate, after consultation with other departments and agencies, through periodic republication by the Secretary, HHS, in the Federal Register. A copy of the list is available from the Office for Protection from Research Risks, National Institutes of Health, DHHS, Bethesda, Maryland 20892.
(b) An IRB may use the expedited review procedure to review either or both of the following:

(1) some or all of the research appearing on the list and found by the reviewer(s) to involve no more than minimal risk,

(2) minor changes in previously approved research during the period (of one year or less) for which approval is authorized. Under an expedited review procedure, the review may be carried out by the IRB chairperson or by one or more experienced reviewers designated by the chairperson from among

members of the IRB. In reviewing the research, the reviewers may exercise all of the authorities of the IRB except that the reviewers may not disapprove the research. A research activity may be disapproved only after review in accordance with the non-expedited procedure set forth in §46.108 (b).
(c) Each IRB which uses an expedited review procedure shall adopt a method for keeping all members advised of research proposals which have been approved under the procedure.
(d) The Department or Agency head may restrict, suspend, terminate, or choose not to authorize an institution's or IRB's use of the expedited review procedure."

JONATHAN MORENO

SECRET STATE EXPERIMENTS AND MEDICAL ETHICS

How does a liberal democratic society facing a national crisis decide what is and what is not permissible with regard to research on human beings—its own people and the citizens of other countries as well?[1] This question is especially apposite in an era of concern over biological and chemical terrorism, but it emerged much earlier and confronted our predecessors on numerous occasions over the past century. While we might imagine that deep concerns about issues of medical ethics are fairly recent—and indeed important dimensions of medical and research ethics have come to light in the past few decades—it is a mistake to underestimate the insights of those who came before us. This chapter demonstrates the history of human subjects research in government contexts, focusing primarily on the United States. History reminds us that vigilant attention to ethical questions is essential. While we have learned much from our predecessors, we should not suppose that later generations, including our own, are any less susceptible to insufficient attention to questions of ethics than our predecessors.

Lessons about the ambiguities of history were brought home to me in 1994 and 1995 when I had the privilege of working as a senior analyst for the President's Advisory Committee on Human Radiation Experiments (ACHRE).[2] In that capacity I was charged with identifying and evaluating the often-secret ethics policies of the United States government during the cold war in the area of radiation research. However, it quickly became apparent that the history of human radiation experiments was closely tied to policies about biological and chemical weapons defense as well. Critical to the story of human subjects research during this era is the fact that, in the years during and immediately following World War II, the United States felt a sense of urgency about protecting liberal democracy. The analogy to our own troubled time is evident, so I will only note in passing that, since September 11, 2001, smallpox vaccine trials with healthy volunteers have been completed in order to rebuild a national stockpile for a suddenly sharpened terrorist threat.[3]

[1] For an account of ethical problems in offshore human subjects research, see the essay by Marcia Angell in this volume, entitled "Cross-Cultural Considerations and Medical Ethics: The Case of Human Subjects Research." *Ed.*

[2] A fuller account of the interplay between government and military policy and human experimentation can be found in Moreno, *Undue Risk*.

[3] "Volunteers Line Up to Test Smallpox Protection," *New York Times*.

A. W. Galston and C. Z. Peppard (eds.), Expanding Horizons in Bioethics, 59-69.

I. THE BEGINNINGS

The story of state-sponsored human experimentation through the lens of human subjects' consent might be said to begin at the University of Virginia. In 1868 the University's most famous medical school graduate, Walter Reed, completed his studies at the age of sixteen. He went on to become the country's most important expert on infectious disease while pursing his army career and directing an important research laboratory in Baltimore. By 1900, a new American army detachment in Havana faced a grave crisis: a pandemic of yellow fever. Since the United States had just gained a foothold in Cuba following the Spanish-American war, the death toll associated with yellow fever posed a threat to national security. Walter Reed was assigned the responsibility of stemming the spread of the disease.

At the time, a popular theory held that the vector of yellow fever was the female silver-backed mosquito. Several of Reed's medical colleagues volunteered to "take the bite" in an effort to assess this theory, and hence to help ameliorate the pandemic. Two of these colleagues, a young doctor and a nurse, died of their exposure; this convinced Reed that in light of his advanced age (he was then forty-nine) he was a poor candidate for the experiment. Instead, to confirm the mosquito hypothesis in a larger population, Reed recruited slightly over two dozen American soldiers and local Cuban workers to participate in a yellow fever experiment. Although human experiments were not unusual in those days, Reed introduced a novel element: He required each of these soldiers and workers who volunteered for the experiment to sign a contract that warned them of the risks of what they were about to do and specified that it was being done freely. The contract was translated into Spanish for the Cuban workers. Not to be denied is the fact that the volunteers were also offered compensation in gold, though it seems that the soldier participants declined this opportunity. The motivation for Reed's innovative contract seems to have been a concern for propriety and immunity from criticism by the prominent Johns Hopkins medical professor William Osler. Reed and the U.S. Army Surgeon General were perhaps concerned that Osler, who had already attacked the ethics of an Italian who used human beings in yellow fever experiments, would turn his critical lens to them. They may also have been aware of a scandal in Prussia about a syphilis experiment that led to a government mandate against all human experiments that failed to attain prior consent from the subjects. Additionally, some United States Congressmen were interested in the issue of human experimentation and introduced a bill to constrain certain kinds of experiments in the District of Columbia.[4]

In addition, early twentieth century advances in medicine were accompanied by political and legal scrutiny about the way these advances were being applied by the medical profession (a pattern that continues in our own time). For example, courts were ruling in favor of patients who were subjected to surgical procedures such as hysterectomies without adequate consent.[5]

[4] Lederer and Grodin.

[5] Katz.

Nonetheless, Reed's military research led to a medical breakthrough—the identification of the mosquito as the vector of yellow fever—as well as a fabulously successful public health intervention. The fact that this occurred within a research framework that was respectful of human beings destined Reed and his team to hold a distinguished place in the history of medicine and medical ethics.[6] In spite of previous cultural reservations about human experiments (recall Mary Shelley's *Frankenstein* and George Bernard Shaw's term "human guinea pig"), the Reed adventure gave both medical and moral validity to the practice. To be sure, scandals about experiments both preceded and succeeded the Reed experience. By the late nineteenth century there were increasing public concerns about the ethics of human experiments and, as I have noted, early attempts at regulation. On the whole, however, the Reed experiment and its proto-consent forms showed that medicine, particularly military medicine, was compatible with medical ethics.

Walter Reed became a cultural hero (aided perhaps by his untimely death while on another assignment a few years later). His experiment largely "inoculated" both medical experimentation and military medicine from association with scandal. This happy condition lasted about fifty years, until the horrendous World War II experiments by German doctors in Nazi concentration camps were revealed.

II. NAZI DOCTORS AND THE NUREMBERG CODE

Following the discoveries of horrific abuses and human experimentation in Nazi concentration camps, many German officials were charged, tried and convicted at the Nuremberg war crimes tribunal. At the behest of the United States, West Germany focused on these experiments in the second of the thirteen war crimes trials at Nuremberg. Twenty-three defendants were implicated, including Hitler's personal physician, Karl Brandt.

Brandt was among the defendants charged with planning or implementing several kinds of experiments, sponsored by the German military, which sought additional information for military pursuits. For example, during World War II the Luftwaffe (German air force) wanted to improve the treatment of hypothermia for its pilots, and they also wanted to know the altitude at which it was safe for pilots to eject themselves from damaged aircraft. Experiments involving freezing and explosive decompression, and many others aimed at benefiting the Nazi regime's war effort, were conducted at the concentration camps. Another category of experiments simulated battlefield injuries: experimental surgeries attempted, e.g., to transplant portions of bone. As many as one hundred thousand people died as the result of these and other experiments in Nazi Germany. The human suffering caused by these experiments defies my limited powers of expression.

Seven of the Nuremberg defendants were hanged for their crimes, and another eight were sentenced to lengthy prison terms. Their Japanese counterparts went

[6] For more information, see Lederer, *Subjected to Science*.

unpunished, even though at least ten thousand innocents died in biological and chemical experiments conducted by the Imperial Japanese army in Manchuria.[7]

However, the prosecution of the Germans did not run a smooth course. The Nazi defense lawyers were able to demonstrate that the World War II allied forces had themselves conducted experiments on people who might well be regarded as vulnerable to coercion. This refutation was surprisingly effective. For example, the defense introduced into evidence a *Life* magazine article from June 1945, which depicted a United States malaria experiment conducted with eight hundred federal prisoners who claimed to have volunteered for the experiment. Although (unlike the Nazi experiments) no one died of malaria and, in my view, these were genuine volunteers who were not facing sub-human conditions and the likelihood of extermination, it became clear that universally recognized rules of human experimentation did not exist. If there were no guidelines and if the allies had conducted similar experiments on imprisoned persons, the Germans' lawyers argued, how could the Nazis be held to a higher standard?

Using this logic, the German lawyers forestalled conviction of the Nazi officials until, in the end, guilty verdicts were based upon the grounds of murder, conspiracy to commit murder, and the new charge of crimes against humanity. The Nazis were *not* found guilty based on the Nuremberg Code, for this would have been *ex post facto* justice. The Nazi defense of human experimentation succeeded in slowing the proceedings, and they also effectively raised the pressing question of human experimentation and coercion. Indeed, the issue of human experimentation was not lost on the three judges, all of whom were American. They understood that international medical ethics standards urgently required articulation. Hence they devoted a portion of their ruling to a ten-point statement that has come to be known as the Nuremberg Code.

With the help of American medical advisors, the judges crafted a document that begins with the phrase "The voluntary consent of the human subject is absolutely essential" to experiments on humans.[8] Other points include the necessity of prior animal experimentation, the elimination of any undue risk, and the right of the subject to terminate his or her involvement in the experiment at any time. Finally, the document states that the medical scientist's responsibility for the well-being of research subjects cannot be transferred to someone else.

III. THE ATOMIC ENERGY COMMISSION

The legal proceedings leading to the Nuremberg Code had a far-reaching effect, even before the trials were complete. In particular, the American Medical Association (AMA) felt compelled to develop its first specific guidelines on medical experiments, anticipating that international standards would indeed be an issue at the Nazi doctors' trial. Andrew Ivy composed the AMA's ethics code prior to the

[7] For more information, see Moreno, chapter four, "Deals with Devils."
[8] *United States v. Karl Brandt et al.,* Vol. II, pp. 181-185. Cited in Moreno, 80.

conclusion of the Nuremberg trials; he was later interviewed during the trials as a key witness, and as a medical advisor to the Nuremberg judges his wording strongly influenced the content of the Code.

The U.S. government also took note of the proceedings. In late 1946 the new Atomic Energy Commission (AEC), headed by MIT engineer Carroll Wilson, was confronted with an experiment that was, unlike the malaria study, a closely held secret and one that would surely have fueled the Nazi defense case.

The AEC administrators discovered, in materials received from the Manhattan Project, that in 1945 seventeen hospitalized patients had been injected with plutonium, apparently without their consent. The scientific purpose of the experiment was to ascertain the human excretion rate of plutonium—a matter of grave concern at the time, since young physicists were exposed to quantities of plutonium daily in the course of developing the mechanisms to release atomic energy. In 1947 a decision was made to maintain the secrecy of the injections, despite repeated requests for declassification of the material by one of the scientists, in order to avoid embarrassment to the government and potential lawsuits.[9]

However, the AEC also decided that it needed to develop a policy to avoid future misuse of its radionuclides, particularly as it was interested in making them available to qualified medical researchers. The implications of the Nazi doctors' trial constituted part of the AEC's concern. Hence, a key element of the resulting policy, as written in a 1947 letter from Carroll Wilson, was that "informed consent" was to be elicited from the patient. The conditions, prohibiting use of a known or suspected harmful substance, are worth quoting in full:

> (a) that a reasonable hope exists that the administration of such a substance will improve the condition of the patient, (b) that the patient give his complete and informed consent in writing, and (c) that the responsible next of kin give in writing a similarly complete and informed consent, revocable at any time during the course of such treatment.[10]

This seems to be the first time that the term "informed consent" appeared, as well as the first time that the standard of family member's consent was established (though that guideline even today seems excessive). Importantly, the AEC established that radiation experiments were to be limited to *therapeutic* testing: experiments must have the possibility of benefiting the patient, not just society at large.

If the AEC letters were meant to be taken seriously, this was surely a dazzlingly inadequate policy. The letters delineated neither a systematic articulation of the policy nor an oversight mechanism. Good intentions notwithstanding, there is reason to believe that the AEC rules were forgotten as soon as the ink was dry. For example, within several years of the statement about informed consent, the AEC, the Massachusetts Institute of Technology, and the Quaker Oats Company implemented a study at the Fernald School in Massachusetts. The Fernald School was a residential institution for young people with a range of objectionable behaviors. The purpose of the study was to determine that the healthful quality of the cereal was as advertised

[9] For more information, see Moreno, chapter five, "The Radiation Experiments."

[10] Letter from Carroll Wilson to Robert Stone, 5 November 1947. Cited in Moreno, 141.

and, additionally, that this brand of cereal was superior to its competitors. The methods of the experiment were highly problematic: radiolabeled nutrients were placed in the students' breakfast cereal. Although permission was sought from the parents, the relevant documents mentioned only that the children had been selected for a "science club" with certain special privileges.[11] The experimenters neglected to mention the trace levels of radiation in the children's breakfast cereal. Throughout the 1950s the AEC participated in a number of other radiation experiments that, at the very least, do not seem to have been compatible with the 1947 directives.

IV. BEYOND THE ATOMIC ENERGY COMMISSION

The AEC's role in the human experiments issue extended to its relationship with other national security agencies. Between 1948 and 1951, an intense discussion took place in an inter-agency committee of the AEC and the Department of Defense (DoD). This joint committee, called "Nuclear Energy for the Propulsion of Aircraft," debated the risks of radiation to the crew of a nuclear-powered aircraft. One effusive proponent who seemed only concerned with short-term results (as opposed to potential long-term effects) was Robert Stone. He claimed that "undetectable genetic effects" on blood or the lifespan were negated by the positive good that could come from learning more about radiation in the body and applying those lessons to military strategy.[12] Others were less sanguine about such approaches. Various possible groups of experimental subjects, including soldiers and long-term prisoners, were suggested and rejected on moral grounds. This process flowed into another fascinating debate within the DoD itself, which was concerned about the authority to conduct defensive atomic, biological and chemical weapons experiments.

The DoD feared that the Soviet regime was unfettered by ethical concerns in human experimentation, particularly when it came to unconventional weapons development. During its complex post-war reorganization, the DoD found that the Pentagon had no policy to govern human experiments. After much discussion, conducted under rules of secrecy, various Pentagon committees could not come to a consensus in favor of an ethics policy. In general, the officers and physician advisors on these panels believed that traditional rules of medical ethics, largely unwritten, had served the purpose of protecting human beings well enough. There was thus some reluctance to establish a written policy that would be subject to legal interpretation. Further, those opposed to human experimentation had a difficult time refuting the claim that human experimentation carried any different risk than normal military service, though some maintained that "it's not very long since we got

[11] For more information, see Moreno, 213-219.

[12] Robert S. Stone, "Irradiation of Human Subjects as a Medical Experiment," January 31, 1950. This document is included in the records of the Advisory Committee on Human Radiation Experiments, Record Group #220, National Archives and Record Administration, Washington, D.C. Cited in Moreno, 145.

through trying Germans for doing exactly the same thing [i.e., human experimentation]."[13]

Rampant offenses implicating the government occurred during this time: in January 1953, a New York City tennis pro named Harold Blaur died in a mescaline-derivative study at the New York Psychiatric Institute, where he was being treated for clinical depression. The study was sponsored by the Army Chemical Corps, which participated in a cover-up with New York State. This event underscored the need for a policy on human experimentation, but the general issue was controversial enough to discourage President Truman's defense secretary from resolving the matter. However, immediately following his confirmation, President Eisenhower's Secretary of Defense Charles Wilson, acting on the advice of the Pentagon's legal counsel, adopted the Nuremberg Code *verbatim* as its policy to govern defensive experiments on atomic, biological and chemical warfare. The policy also included a written consent requirement. This directive was signed on February 26, 1953, and established the Nuremberg Code as the Pentagon's human experiments policy. It was the first and last time that the code was adopted by a government entity as official policy. Unfortunately, the document was also classified as top secret.

Not surprisingly, given the top-secret status of the adoption of the Nuremberg Code, the results of this policy were mixed. In 1954, about a year after the policy was adopted, an air force officer in charge of flash-blindness studies complained to superiors that he had heard only rumors about a new rule for human experiments. He claimed that "no serious attempt has been made to disseminate the information to those experimenters who had a definite need-to-know."[14] It seems that several of the men in the study had experienced temporary blindness, which elevated this official's concern about his ignorance of the specific policy. Clearly, the top-secret nature of the original document made the policy difficult to disseminate.[15]

The disappointing story of the actual effect of the DoD's Nuremberg Code-based policy is, in my view, attributable to numerous cultural factors that characterized both the military establishment and the medical profession in the 1950s. Simply put, neither the military nor the medical profession was prepared to embrace a notion of written informed consent, nor was it clear to the protagonists exactly what that meant in practice. While a handful of human testing projects seem to have been accompanied by written consent (e.g., panic studies stemming from fear of radiation at atomic test shots), many others were not. Examples of the latter include the over two hundred thousand men who were deployed at Camp Desert Rock in Nevada for

[13] Advisory Committee on Human Radiation experiments. Cited in Moreno, 148.

[14] Advisory Committee on Human Radiation Experiments. Cited in Moreno, 292.

[15] Parenthetically, no similar policy seems to have been adopted by the CIA, which was actively engaged in human experiments in the 1950s. For example, the infamous and legendary MKULTRA project was one of a number of activities intended to apply biological and chemical agents for undercover espionage and sabotage. One aspect of the project was an extensive network of financial support for LSD and behavioral experiments, often using "front" foundations and corporate entities to distribute funding. One victim of the project was a CIA scientist assigned to Fort Detrick in Maryland, Frank Olson, whose expertise was anthrax. Olson met his death when he tumbled from a New York City hotel room window in 1953. When details of his role were revealed in the 1970s, the government's explanation of his death was a psychotic episode brought on by LSD dosing, but subsequent forensic investigations arranged by his family point instead to an assassination.

above-ground nuclear tests, the so-called "atomic soldiers." This activity was classified as training and indoctrination rather than as a medical experiment.[16] In a very different example, soldiers who were exposed to LSD (lysergic acid diethylamide) in the 1960s seem to have known they were going to be given a hallucinogen, but they were not aware of other details, such as when the exposure would take place, what quantity of LSD they would be given, or what the goals of the study were.[17] On the other hand, the infamous case of Dr. Frank Olson (a doctor who was administered LSD and later appeared to have killed himself; the case was declassified in 1976) indicates that LSD trials were applied to unsuspecting persons who, as they were unaware of the LSD trials, were clearly unable to give consent.[18]

In 1975 the Army Inspector General reflected on twenty years of experience with the Nuremberg Code policy and concluded that there was a "startling lack of consistency" in interpretation of this policy.[19] Further, when one of the soldiers affected by military medical experimentation attempted to sue the federal government for injuries sustained in the LSD experiments, the U.S. Supreme Court ultimately ruled against him. The Supreme Court, in *Thornwell v. The United States of America,* cited the doctrine that a member of the armed forces cannot sue for harms incurred in the line of duty.[20] The question, of course, is whether participation in human experimentation qualifies as "the line of duty."[21] Issues of informed consent and coercion arise in this context. This is particularly true since it seems that the Nuremberg-based policy was not systematically understood, much less implemented. I believe that the five-member Supreme Court majority gravely erred in this decision and, as a result, did substantial damage to the morale of members of the armed forces.

It seems clear that the DoD policy generally struggled. Yet, oddly enough, there was an apparently successful and highly publicized human experiments program at Fort Detrick from 1953 to 1974, where over two thousand soldiers who were members of the Seventh Day Adventist Church volunteered for assignment in Operation Whitecoat. These men participated in numerous studies involving potential biowarfare agents. Follow-up studies on two hundred veteran volunteers in 1998 did not suggest that the men suffered ill effects (though it is feasible that this group of volunteers is not a representative sample).

Operation Whitecoat was terminated in 1974 as part of Nixon's directive to shut down the biological weapons program at Fort Detrick. However, the infectious disease institute (U.S. Army Medical Research Institute of Infectious Diseases) was not shut down but charged with developing medical treatment for dangerous contagious diseases. The cessation of Operation Whitecoat effectively eliminated the source of human subjects volunteers. Gradually the Medical Research Volunteer

[16] Moreno, 206.
[17] Ibid., 251-254.
[18] Ibid., 191-192.
[19] Ibid., 180.
[20] *Thornwell v. The United States of America*, 471 F. Supp. 344.
[21] The details of this case are particularly disturbing. James Thornwell, while under investigation in France for an alleged document theft, was kept in solitary confinement and was refused meals and sleep before administered LSD to test its efficacy as a "truth serum."

Subjects (MRVS) program was born, with participants recruited from groups of medics. The medics were assigned to Fort Detrick's infectious disease institute with the understanding that, in addition to regular laboratory duties, they would make themselves available to volunteer as research subjects. I conducted interviews with MRVS volunteers, and these exchanges convinced me that the reserve volunteers were a highly motivated and well-informed group. Some had not volunteered for an experiment in nearly three years, while others had volunteered for many. Two features of the MRVS program stand out. First, the group had a representative on the ethics committee. Second, MRVS members were not allowed to receive remuneration for research participation; monetary incentive was therefore not an issue (unlike volunteers in various civilian locales, for whom payment might be decidedly coercive). It is fascinating and ironic that, following a decidedly mixed history of human subjects protections, the Army had created one of the most admirably ethical human research programs in the world.

V. CONCLUSION

The mid-1970s was a critical period in the story of research protections. It was then that the Army and CIA hallucinogenic experiments were revealed and the original Nuremberg Code-based DoD policy memorandum was fully declassified and the policy itself analyzed. Several individual cases were acknowledged, including those of Blaur and Olson, and compensation arrangements were made. But as dramatic as these revelations were, they were somewhat overshadowed by contemporary scandals in civilian medical experimentation, such as the Tuskegee Syphilis Study and the gradual shutdown of prison experimentation.

How do we explain the fact that the infamous Tuskegee Syphilis Study and other events in the civilian medical world have acquired a familiar role in bioethics scholarship, but the national security experiments have received only minimal attention? I theorize that a lack of familiarity with the national security history and context is the culprit, including both the military and foreign policy aspects. Of course, little historical investigation and reflection is possible on such matters until the declassification of various documents. In addition, the Nixon administration's dismantling of the U.S. biological and chemical weapons programs removed this research area from the lens of inquiry just as bioethical debates were emerging on topics such as human experimentation. Indicative of this is the fact that the Nuremberg Code has long been recognized as a crucial document in research ethics, but its application and relevance to United States military research have remained oblique until very recently. Only now is this lacuna in the scholarship being corrected. As our knowledge and awareness of the history of human experimentation and human research protocols increases, we can no longer ignore the importance of military medical experiments in the bioethics corpus.

These issues are not outdated. The fine line between "the line of duty" and "human subjects experimentation" in the military context arose again during the 1991 Gulf War. During Operation Desert Storm, informed consent requirements were suspended for several chemical and biological agents that were used by soldiers in Desert Storm.[22] At that time, few bioethicists commented on this transgression. Allegations about Iraq's tests of chemical and biological agents on political prisoners resurfaced in 1998, when U.N. inspectors reported that they were about to encounter a paper trail documenting the use of prisoners in experiments by the regime of Saddam Hussein. In the events following September 11, 2001 and the expulsion of Saddam Hussein from Iraq, questions of national security and medical ethics have burst onto the national scene with fervor. One issue is the need for government approval of potential bioweapons therapies that cannot ethically be tested in humans, yet may have to be given to large numbers of Americans in the event of a widespread biological attack. Another issue is the testing of vaccines to prevent a catastrophic disease event among civilians (such as a recurrence of smallpox). It remains to be seen whether the field of bioethics will achieve the imaginative leap to critically assess and integrate this neglected area. If it does not, it will fail to provide a relevant critical voice to the confluence between modern politics and scientific weapons research.

[22] See Moreno, chapter nine, "Once More Into the Gulf."

REFERENCES

Katz, Jay. "Informed Consent—A Fairy Tale? Law's Vision." *University of Pittsburgh Law Review* 29 (Winter 1977): 137-174.

Lederer, S. *Subjected to Science: Human Experimentation in America Before the Second World War*. Baltimore: Johns Hopkins University Press, 1995.

Lederer, S. and M.A. Grodin. "Historical Overview: Pediatric Experimentation." In *Children as Research Subjects: Science, Ethics and Law*. Edited by Michael A. Grodin and Leonard Glantz. New York: Oxford University Press, 1994.

Moreno, J. *Undue Risk: Secret State Experiments on Humans*. New York: Routledge, 2003.

Stone, R.S. "Irradiation of Human Subjects as a Medical Experiment." In the records of the Advisory Committee on Human Radiation Experiments, Record Group #220, National Archives and Record Administration, Washington, D.C.

Thornwell v. The United States of America, 471 F. Supp. 344.

United States v. Karl Brandt et al., Vol. II, pp. 181-185.

"Volunteers Line Up to Test Smallpox Protection." *New York Times.* November 3, 2001.

Wilson, C. Letter to Robert Stone, 5 November 1947. ACHRE No. DOE-052295-A-1.

MARCIA ANGELL

CROSS-CULTURAL CONSIDERATIONS IN MEDICAL ETHICS

The Case of Human Subjects Research

For the past decade or so, one of the most contentious issues in medical ethics has been the question of whether standards for human research should be the same across cultures.[1] The issue became public in the late 1990s when a heated debate arose about the use of placebos in clinical trials in underdeveloped countries.

A clinical trial is a method for testing a new treatment for a disease by comparing it with the old treatment in volunteers to see which is better. The volunteers are randomly divided into two groups – an experimental group and a control group. The experimental group receives the new treatment, and the control group receives the old one. If there is no known effective treatment, then the control group may receive a placebo (sometimes called a sugar pill).

This chapter will discuss the appropriateness of placebo-based clinical trials for testing HIV medications in pregnant women in Africa. I begin with some general comments about the aims of science in light of ethical imperatives about research in humans. I then discuss the particularities of placebo-controlled anti-HIV drug trials and argue that such trials are unethical. I also contend that research on human subjects is best conducted within the sponsoring country and, further, that in all cases researchers have a responsibility not to leave any human subjects with deadly diseases untreated if they have the means to treat them.

SCIENCE VERSUS ETHICS

Clinical trials raise formidable ethical problems that stem from the responsibilities of researchers to protect human subjects and also to advance the interests of science. It would be nice if these dual responsibilities always coincided, but unfortunately, they may not. On the contrary, there can be an inherent tension between the search for scientific answers and concern for the rights and welfare of human subjects enrolled in that endeavor.

As a hypothetical example, consider the problem of researchers who want to test a new vaccine against HIV infection. Scientifically, the best way to proceed would be to choose healthy participants, give the vaccine to the experimental group and a

[1] Though the question of experimentation on humans has a robust history in the United States (see J. Moreno's chapter in this volume), the issue of offshore research trials emerged most forcefully in the late 1990s. *Ed.*

A. W. Galston and C. Z. Peppard (eds.), Expanding Horizons in Bioethics, 71-84.

placebo to the control group, and then inject both groups with HIV and compare the infection rates. If infection occurred mainly in the control group, but not in the vaccine group, it would prove that the vaccine worked. Such a trial would be simple, fast, and conclusive. In short order, we would have a clear answer to the question of whether the vaccine worked. Further, if the answer is affirmative, this vaccine might have enormous public health importance and save millions of lives.

Yet almost everyone would agree that such a trial would be unethical. If asked why, most people would probably suggest that human research subjects should not be treated as guinea pigs—that is, they should not be used merely as a means to an end, particularly when the "means" involve suffering or death. There would be an instinctive revulsion against deliberately infecting human subjects with a lethal disease, no matter how important the scientific question.

Researchers in this hypothetical example would have to make scientific concessions for ethical reasons. Since researchers would be prohibited from injecting HIV, they would simply have to wait to see how many people in each group (vaccinated and unvaccinated) became infected over the natural course of their lives. That could take many years, a very large number of subjects, and meticulous monitoring. Even if researchers chose subjects at high risk of becoming infected with HIV—say, intravenous drug users—the research subjects would have to be followed for a long time to accumulate the number of infections needed to permit a statistically valid comparison between the vaccinated and unvaccinated groups. Even then, the results might be hard to interpret, because of possible but unknown differences in the exposure between the two groups. In short, doing an ethical trial would be far less efficient and conclusive—and much more expensive—than simply injecting the virus into research subjects.

Conducting the trial ethically and more slowly might have repercussions beyond a loss of scientific efficiency. If the vaccine turned out to work, a long, laborious and inconclusive (but ethical) clinical trial could also mean a loss of lives—the lives of all those throughout the world who contracted HIV infection for want of a vaccine during the extra time it took to do an ethical trial. There would have been a trade-off between the welfare of participants in the trial and the welfare of the far larger number of people who would benefit from finding an effective vaccine quickly. Thus, either human subjects would suffer by being deliberately exposed to HIV infection in an unethical trial; or (if the vaccine worked) future patients would suffer by having been deprived of a vaccine while an ethical, but longer, trial was conducted.

This hypothetical example of the tension between science and society, on the one hand, and ethics, on the other, is admittedly extreme. Nearly everyone would agree on the right course of action in this case, regardless of the utilitarian claim that injecting human subjects with HIV would do the greatest good for the greatest number. Yet there have been many real experiments involving no less extreme choices in which researchers sacrificed the welfare of human subjects to the interests of science and future patients, and believed they were right to do so.

ETHICAL PRINCIPLES GOVERNING RESEARCH ON HUMANS

Recognition of the frequent tension between science and ethics has given rise over the years to certain principles that govern clinical research on human subjects. This trend was initially encapsulated in the Nuremberg Code after the Second World War. Several tenets bear articulation. First, researchers should hold the welfare of their human subjects above the interests of science and society. When they enlist volunteers for clinical trials, researchers should assume responsibility for those volunteers, just as clinicians do when they assume the care of patients. The research should not expose volunteers to risks of grievous harm, and all risks should be minimized. One of the consequences of this principle is that researchers should be genuinely uncertain before a trial begins as to whether the experimental treatment will turn out to be better than the control treatment.[2] This state of uncertainty has been termed "equipoise."[3] Equipoise requires that control subjects not be given placebos if there is a known effective treatment. New treatments should be tested against effective old ones, if such exist, not against placebos. If they already know that the experimental treatment is better (and presumably just want to find out how *much* better), then they are guilty of deliberately providing their control group with inferior treatment. If placebos are used in this situation, then the control subjects are being given inferior treatment—namely, nothing.

Second, participation in clinical trials should be entirely voluntary. Researchers must explain the nature of the research to potential volunteers before enrolling them in trials and obtain their "informed consent." The explanation should include all the salient risks and benefits of participation. Potential volunteers must then have the opportunity to consent or refuse to participate without penalties or undue inducements. That is, there should be neither bribery (as would be the case if payments for participating were excessive) nor blackmail (as would be the case if general medical care were withdrawn from those who refused to participate).

These two principles have been incorporated in a number of national and international codes of ethics, as well as in legislation and regulations in some countries. The most celebrated of these was the Nuremberg Code of 1947.[4] It was the aftermath of the notorious Nazi Doctors trial, in which German doctors were found guilty of crimes against humanity for their brutal mistreatment of human subjects. The Nuremberg Code emphasized the necessity of voluntary consent, but also stipulated that there must be no *a priori* reason to believe that death or disabling injury would result from the research.

Perhaps the most important of the subsequent ethical codes is the Declaration of Helsinki, first promulgated by the World Medical Association in 1964 and revised several times since then, most recently in 2000.[5] It states, "In medical research on human subjects, considerations related to the well-being of the human subject should take precedence over the interests of society." It explicitly supports the concept of

[2] Angell, "Patients' preferences."
[3] Freedman, 141-5.
[4] Annas and Grodin.
[5] World Medical Association, "Declaration of Helsinki."

equipoise, including a prohibition against researchers treating control groups with placebos if there is known to be an effective treatment:

> The benefits, risks, burdens and effectiveness of a new method should be tested against those of the best current prophylactic, diagnostic, and therapeutic methods. This does not exclude the use of placebo, or no treatment, in studies where no proven prophylactic, diagnostic or therapeutic method exists.[6]

In the United States, the Department of Health and Human Services promulgated regulations in 1991 (now known as the Common Rule) that govern most, but not all, clinical research sponsored or regulated by U.S. agencies. These regulations are less absolute than the Nuremberg Code and the Declaration of Helsinki, but they also emphasize the importance of minimizing risks to human subjects.[7] The regulations provide for the review of clinical research by institutional review boards (IRBs)—formerly called ethics committees. These are usually constituted by the institutions sponsoring the research. Their function is mainly to ensure that the research conforms to the two principles described above.[8]

THE AZT TRIALS IN HIV-INFECTED PREGNANT WOMEN

In recent years, it has become recognized that research conducted in underdeveloped countries by researchers from developed countries can pose special problems. These were brought to public attention in 1997, when it became known that the U.S. government was sponsoring numerous trials of AZT (zidovudine) treatment for HIV infection in underdeveloped countries. In these experiments, some pregnant women were given placebos instead of known effective treatments. This was done even though it was understood that thousands of infants would contract an avoidable lethal disease.

About a quarter of HIV-infected women are known to transmit the disease to their offspring, which is almost always fatal for these children. In 1994 it was shown that giving women AZT during pregnancy and delivery would greatly reduce the transmission rate—from about 26 to 8 percent.[9] Almost immediately after publication of this research, the U.S. Public Health Service recommended that all HIV-infected pregnant women receive this intensive course of AZT—known as the "076 regimen." There was no question that it would save many lives. (This regimen has since been superseded by even more effective regimens consisting of multiple antiretroviral drugs.)

But there was a problem. The drug was extremely expensive, and many pregnant women could not afford it. That was particularly true in underdeveloped countries, precisely where prevention was most needed. In Africa, millions of pregnant women were infected with HIV. The entire health expenditure in Malawi,

[6] Ibid.

[7] Department of Health and Human Services, "Protection of Human Subjects."

[8] See C.Kristina Gunsalus' chapter in this volume for current questions on the expansion of IRBs in academic contexts. *Ed.*

[9] Connor et al.

for example was at that time only about a dollar a person; yet the 076 regimen was priced at about $800 per pregnancy Furthermore, the regimen was difficult to administer. It required intravenous treatment during delivery, and long-term oral treatment throughout pregnancy and for six weeks in the newborn—a difficult proposal in a country with limited medical and financial resources.

There was therefore intense interest in trying to find a shorter, cheaper, and easier regimen of AZT that might work as well as the 076 regimen. There were reasons to be optimistic about this possibility. The original study of the 076 regimen had shown that women who started the drug late in their pregnancy were no more likely to transmit the infection than those who started early. Other evidence suggested that the most likely time of HIV transmission was right at the time of birth. Studies had also shown that the oral form of the drug got to the blood in high levels. There was every reason to hope, therefore, that administering AZT orally near the time of delivery would be effective.[10] Starting in 1994, soon after publication of the 076 study, the U.S. government launched ten clinical trials of shorter, cheaper regimens of AZT in underdeveloped countries—mainly in Africa. No one objected to the goal of trying to find a less intensive AZT regimen that might be effective and would enable more women to access treatment.

CRITICISMS OF THE TRIALS

The trouble was that in all but one of these trials, instead of using the 076 regimen in the control groups to see if the short regimens worked as well, the researchers used placebos as the comparison. That meant that about 26 percent of babies born to women in the control groups would develop AIDS, as opposed to about 8 percent if the 076 regimen had been used. In other words, the researchers had deliberately allowed thousands of infants in their care to contract preventable HIV infection by failing to offer them a proven treatment.

These clinical trials had been going on for about three years, when, in the fall of 1997, Drs. Peter Lurie and Sidney M. Wolfe of Public Citizen's Health Research Group published a highly critical article in the *New England Journal of Medicine* (*NEJM*).[11] Lurie and Wolfe argued that it was fine to look for shorter regimens, but that new treatments should be compared with the full 076 regimen, not with placebos. In the U.S., they said, subjects in the control group would have received the best standard treatment. In fact, trials would not be approved unless they did. Human subjects in underdeveloped countries, they said, should be treated with the 076 regimen as well.

At that time I was executive editor of the *NEJM*, and I wrote an editorial in the same issue supporting the position of Lurie and Wolfe.[12] I pointed out that the AZT trials were not unique in using underdeveloped regions of the world for research that would not be approved in the U.S. Placebo-controlled trials of treatments for other

[10] Lurie and Wolfe.

[11] Ibid.

[12] Angell, "Ethics of Clinical Research," 847-9.

diseases were being conducted in Africa, despite the fact that there already were known effective treatments. I argued that this was a violation of the Declaration of Helsinki, which required that the control group receive the best available treatment: "The benefits, risks, burdens and effectiveness of a new method *should be tested against those of the best current prophylactic, diagnostic, and therapeutic methods.*"[13] Since AZT was known to be highly effective in stopping mother-to-infant transmission of HIV, the researchers were permitting many infants in their care to be born with preventable HIV infection. I compared the practice to the Tuskegee Study, in which the U.S. Public Health Service denied treatment for syphilis to a group of black men in rural Alabama.[14] In that study, the men were observed over many years without treatment, even after penicillin had become available. What was the difference, I asked, between doing that and observing untreated HIV-infected pregnant women in Africa? It was wrong, I asserted, to use powerless people in underdeveloped countries to do research that could not be done in the affluent parts of the world simply because researchers could get away with it.

THE RESPONSE

The reaction was immediate. There were multiple stories in the media about the controversy.[15] Many people in the HIV research community heatedly justified the trials. They were particularly indignant about the comparison with the infamous Tuskegee study. Two members of the *NEJM's* editorial board resigned in protest, and there were calls for my removal as executive editor. The Directors of the National Institutes of Health (NIH) and Centers for Disease Control (CDC)—the governmental agencies that sponsored the research—defended the trials in the *NEJM* a few weeks later.[16] According to them, the fact that the studies would be unethical in the U.S. (which they did not deny) did not mean that the trials were unethical in Africa. Most of all, they were offended by the comparison with Tuskegee. They offered four main arguments.

First, they pointed out that pregnant women in Africa would probably not receive the 076 regimen outside the trials, because it was prohibitively expensive. In their view, that meant there was no ethical requirement to offer it within the trials. In short, they argued that the women in the control groups were no worse off than they would have been if they were not in the trials. Second, they said that the relevant question in Africa was not whether a short regimen was as good as the 076 regimen, but whether it was better than nothing. After all, these researchers argued, it was unlikely that the 076 regimen would ever be affordable there, so why bother testing it in a comparison group? Third, they claimed that it was important to do the research in Africa, because that is where the results would be applied. People in underdeveloped countries were in desperate need of an affordable regimen to

[13] World Medical Association, "Declaration of Helsinki," emphasis added.
[14] Angell, "Tuskegee revisited."
[15] French, 1.
[16] Varmus and Satcher.

prevent HIV transmission from mother to child, and that is exactly what would emerge from this research. As they put it, "A placebo-controlled trial may be the only way to obtain an answer that is ultimately useful to people in similar circumstances."[17] Finally, they argued that the trials were justified because local authorities had approved them. In their words, the NIH and CDC "decisions to support these trials rest heavily on local support and approval."[18]

A CLOSER LOOK AT THE ARGUMENTS

Let's look at these arguments more closely. The first thing to note is the way in which they blur two quite different issues: research in relatively small groups of human subjects for a limited time, and treatment of the entire population for the indefinite future. Thus, the fact that African women generally could not afford AZT treatment was offered as an excuse for their not receiving it in clinical trials—despite the fact that the circumstances were very different. Giving AZT to women in the control groups of the trials would have added very little to the costs of the research. In fact, drug companies usually offer drugs free to researchers. The real costs of trials lie mainly in logistics and personnel. The second thing to note is the way in which the arguments accept the pricing of drugs as something akin to a law of nature: drug prices were high because they had to be. But in fact, brand name drugs are generally priced far above their costs. (In recent years, this fact has been underscored by offers of several drug companies, embarrassed by public disapproval of their efforts to keep generic drugs out of Africa, to reduce by up to 80 percent their prices for anti-HIV drugs in Africa.) Should research be designed to accommodate the peculiarities of patent law and drug pricing?[19] What about the argument that since most African women would not be able to afford AZT anyway, those receiving placebo were no worse off than they would be if they weren't being studied? That is probably true, but it is hardly an ethical position. It is taking advantage of their poverty to exploit them. Indeed, unethical research has historically been "justified" by the claim that the particular human subjects were in a bad situation anyway. In the Nuremberg Trials, some of the Nazi doctors used this "they were doomed, anyway" defense. This same argument was once made in defense of the Tuskegee study. In that study, the U.S. Public Health Service observed 412 black men with latent syphilis to learn about the course of the untreated disease. Some defenders of the study pointed out that black men in rural Alabama would be unlikely to get treated for syphilis anyway, so they were not being denied anything.

On the contrary, it cannot be emphasized enough that researchers should be responsible for the welfare of the human subjects they enlist in their studies. Economic, social, and political conditions in the larger population do not justify exposing subjects to preventable risks. Even though AZT may have been

[17] Ibid.
[18] Ibid.
[19] Loff and Heywood.

prohibitively expensive for African women in general, with little trouble researchers could have offered the 076 regimen to the women in their trials. It is difficult for me to understand how researchers could stand by while infants in their charge contracted HIV infection.

The argument that the only relevant question in Africa was whether the short regimen was better than nothing was also unpersuasive. What we really needed to know as a first step was whether short regimens were as good as the 076 regimen. If they were, or were close enough, there would be no need to go any further. We already knew from earlier research in many parts of the world what the approximate rate of transmission was without treatment, so placebo controls in further research were not necessary. And, for all practical purposes, we could have decided on the basis of this earlier research whether the short regimens were a worthwhile improvement over nothing. In fact, the trial that yielded the most useful information was the one trial that did not employ a placebo control group.[20] These researchers, from the Harvard School of Public Health, insisted that it would be unethical to do so, despite pressure from an NIH advisory group.

The third argument—that the fruits of this research would be used in the countries where the research was conducted—is almost certainly wrong. There have been too many cases in which research in poor countries was of benefit only to wealthy ones. There was no reason to believe that this time it would be different. If the 076 treatment was out of reach of ordinary Africans because of its price tag of $800, so were the short regimens, which would cost perhaps a tenth as much. But there was no question that the U.S. and Europe would welcome cheaper treatments and use them. There were no commitments, either by the sponsors or the drug companies, to provide the fruits of this research to the countries in which it was done.

Finally, the argument that local authorities had approved the trials is irrelevant at best. In most of the countries where this research took place, there were repressive governments that were hardly representative of their people. The authorities who approved the trials were no more representative of the human subjects than officials in the U.S. Public Health Service were of the men in the Tuskegee Study. Further, many people stood to benefit personally from the research. Millions of dollars were poured into the local economies (but not directly to participants), and local doctors were paid for their participation, were often included as co-authors on papers, and were given free trips to meetings in the U.S. It is sometimes said that these trials were approved because of different cultural traditions, but it is more likely that officials and researchers in the host countries shared the goals and mindset of U.S. officials and researchers, and that financial incentives made the research offers hard to refuse.[21]

In 1993, the Council for International Organizations of Medical Sciences (CIOMS), in collaboration with the World Health Organization, specifically

[20] Lallemant et al.

[21] Angell, "Ethical imperialism?"

addressed the issue of clinical trials done in foreign countries.[22] The guidelines were at odds with the later NIH and CDC position. They state:

> An external sponsoring agency should submit the research protocol to ethical and scientific review according to the standards of the country of the sponsoring agency, and the ethical standards applied should be no less exacting than they would be in the case of research carried out in that country.

In other words, if the NIH sponsors a study in, say, Malawi, that study must meet the standards of both the U. S. and Malawi. Protections should be at least as great—and surely not less—than for human subjects in the United States.

The placebo-controlled AZT trials were halted soon after the controversy erupted. Almost immediately, however, there were American-led efforts to revise the Declaration of Helsinki and the CIOMS guidelines to accommodate such trials in the future.[23] They would require that subjects in control groups receive only the usual treatment available in the region, despite the fact that researchers could in most cases easily supply the standard treatment. After much debate, the Declaration of Helsinki was in 2000 strengthened, not weakened. It remains to be seen whether the CIOMS guidelines will be substantively changed.

WHY GO OFFSHORE?

The AZT trials were sponsored by the U.S. government, not by corporate industry. Most clinical trials, however, are sponsored by pharmaceutical companies. Earlier I referred to the massive export of clinical trials from wealthy countries to poor ones. That is mainly a reflection of the rapid increase in clinical trials conducted by pharmaceutical companies in underdeveloped countries. When did this trend begin, and why?

The movement toward offshore research began in 1980, when the U.S. Food and Drug Administration (FDA) first agreed to accept foreign trials as evidence of safety and effectiveness in new drug applications (NDAs). Manufacturers of prescription drugs are required to demonstrate safety and effectiveness before they can get FDA approval to market a new drug. They submit their evidence with the new drug application. Hence, permission to include evidence from foreign clinical research opened new possibilities.

The exportation of clinical trials accelerated during the 1990s because of an explosion in the total number of clinical trials for pharmaceutical products and the resulting scarcity of human subjects. From December 17 through 22, 2000, the *Washington Post* published a six-part exposé of medical experimentation in underdeveloped countries. Called "The Body Hunters," this series of articles documented the wholesale movement of clinical trials overseas as the search for human subjects grew increasingly desperate. Since companies feel under increased commercial pressures to cut corners, over a quarter of new drug applications now rely on foreign research.

[22] Council for International Organizations of Medical Sciences.

[23] Brennan; Levine.

Researchers who submit NDAs to the FDA are registered with the FDA, and most foreign trials include some from the host country. It is illuminating to look at the growth in the number of foreign researchers registered with the FDA. In 1991, according to the *Washington Post*, there were only two researchers from Africa, five from South America, and one from Eastern Europe. By 2000, there were 266 from Africa, 453 from South America, and 429 from Eastern Europe.[24] Although this is an indication of the rapid exportation of clinical research, it is only the tip of the iceberg. Much research in underdeveloped countries is not registered with the FDA or any other body with oversight responsibilities.

Why are companies going offshore? There are several reasons. First and most obviously, it is cheaper and in many respects easier. Some private research companies specialize in conducting trials in underdeveloped areas of the world. Although they pay local doctors far less than they do in the U.S., by local standards the rewards are munificent. For example, the *Washington Post* reported that one doctor in Eastern Europe, who had a monthly salary of only $178, could make up to $2000 for each patient he enrolled in clinical trials. He and his colleagues could hardly afford not to enter the business. Patients, too, are readily enticed by small amounts of money and promises of free care. The labor costs of conducting the trials are also relatively low by Western standards, although they may represent a significant influx for the local economy.

In addition, many underdeveloped countries are governed by authoritarian regimes. These regimes often welcome the influx of U.S. dollars. They also welcome the legitimizing effect of the presence of U.S. researchers, given the esteem in which American research is held. Government officials may then "encourage" people to volunteer. The concept of informed consent becomes cursory and hollow in this context. Whole villages and provinces may be enrolled in trials, sometimes with promises of free medical care.[25] In a few such instances, subjects were later interviewed by reporters. When they were asked whether they understood the nature of the research and that participation was voluntary, it was clear they had not. In fact, the very notion of voluntariness was strange to them. They simply assumed they had to participate, or at least could not change their minds after the research began.

Recently, the Department of Health and Human Services' Office for Human Research Protections announced an investigation of a large study sponsored jointly by the Harvard School of Public Health and Millenium Pharmaceuticals in rural China. The study involved drawing blood from 200,000 peasants to form part of a large genetic database owned by Millenium Pharmaceuticals. Reportedly, peasants hesitant to participate were subjected to "thought work" to persuade them.[26] After the controversy over the AZT trials in Africa erupted, reporters interviewed some of the women who had participated in the research.[27] They found that they did not really understand the nature of the research, nor that participation was voluntary.

[24] Brown.
[25] Chang, D12.
[26] Pomfret and Nelson, A01.
[27] French.

That was also true of the Chinese peasants. In particular, the African women found it difficult to believe that American doctors might actually be giving them sugar pills instead of real treatment. And who can blame them?

Pharmaceutical companies and researchers in offshore trials also have the advantage that people in underdeveloped countries are not likely to be taking drugs that would disqualify them from participating in trials of new drugs. Most clinical trials exclude subjects who have already been treated for the disease in question, as well as subjects who are taking any other drugs. As people in affluent countries are increasingly being medicated for all sorts of minor and major ailments, it is growing difficult to find therapeutically "naïve" human subjects. Almost everyone is taking some pill or another for something. In underdeveloped countries, however, it is possible to find large numbers of people who are taking no prescription drugs whatsoever.

CIRCUMVENTING FDA REGULATIONS

Perhaps the most important reason for going offshore is that it is a way of circumventing FDA regulations. Whereas drug companies are required to submit investigational new drug applications (IND) to the FDA before they can begin clinical testing of new drugs, they do not have to do that for trials conducted abroad. An IND includes a description of the protocol for the research, including plans for IRB (institutional review board) approval and monitoring of progress. The FDA has the opportunity to deny approval or demand changes. That is not the case for foreign trials. The FDA may not be notified of offshore trials until after the trials are completed, when an NDA (new drug application) is submitted. Only then—when there is no longer an opportunity to intervene—does the sponsor attest to the way in which the research was done. Only then is the FDA told whether there was IRB approval and informed consent. That is very late in the game.

For research not directed toward FDA approval, there may be no oversight at all. Companies may conduct preliminary studies on people in underdeveloped countries to work out problems before formal testing even begins. These subjects are quite literally used as guinea pigs, for research that really should be done on experimental animals is done on them. Although some research in the U.S. and other developed nations also escapes formal oversight, there are generally more restrictions on what sponsors can get away with. Furthermore, the FDA almost never conducts on-site inspections abroad. It conducts very few in the U.S., but at least there is the possibility. They simply take the word of sponsors for research done in the underdeveloped countries.

Given the laxity of oversight, it is not surprising that companies may go abroad to conduct studies that they could not do at home. In fact, there have been cases of research that were turned down by the FDA then later done abroad. This is analogous to IRB shopping, in which sponsors go to a more congenial IRB after they have already been turned down by a more demanding one. Here, companies are country-shopping.

SHOULD CLINICAL RESEARCH BE EXPORTED TO UNDERDEVELOPED COUNTRIES?

Some critics are less concerned with the design and conduct of trials in underdeveloped countries than with the whole idea of performing research there in the first place.[28] They believe it is virtually impossible to carry out ethical research in underdeveloped countries because the power relationship is unequal and it is inherently exploitative. Although I would not go that far, I do believe the amount of research sponsored by wealthy countries and carried out in poor ones should be sharply curtailed. Such research is driven too much by the search for profits and by ambition. It offers quick answers precisely because it is so easy to cut corners and avoid difficult ethical questions and time-consuming revisions to research protocols.

It seems to me that before a study is exported to underdeveloped countries, two important questions should be asked. First, would it be possible to do the research in the sponsoring countries? And second, why is it being exported? If it would be possible to do the research in the sponsoring countries, then it should be. There are, unfortunately, plenty of HIV-infected pregnant women in the affluent countries, and the AZT trials should have been done on them. Then there would have been no placebo groups. As was evident with the African women, whatever benefits accrue to human subjects in poor countries by being in trials are probably outweighed by the harms.

It is often implied that research must be done in the neighborhood where the results of research will be applied. Though this argument is theoretically true, it is rarely borne out in practice. Whatever could have been learned from AZT trials in the developed world would be no less likely to be used in Africa than if the research had been done there. The notion that people can benefit only by research done in their neighborhood perpetuates the confusion between research and treatment. The benefits of research can be realized anywhere in the world—no matter where the research is performed.

In my opinion, the only clinical research that needs to be done in underdeveloped countries is research on diseases that uniquely affect them—such as malaria or sleeping sickness or schistosomiasis. To have localized research for trenchant localized problems is ample justification for a certain type of offshore research. I only wish that there were more of this sort of research. The fruits of research on these sorts of diseases would *per force* be used in the regions where the research was conducted, and that is how it should be. Unfortunately, it is not a high priority either for the pharmaceutical industry or the NIH and CDC.

Although there are arguments on both sides of this debate, I believe that on balance, research should not be done in underdeveloped countries unless it concerns diseases that are virtually unique to that part of the world. I also believe that regulations governing research in poor regions of the world should be every bit as stringent—and enforced just as vigilantly—as in wealthy countries. There is no justification for the present situation in which the standards of research and human

[28] Glantz and Grodin.

subjects protection are looser precisely where human subjects are most vulnerable. We cannot continue to treat the underdeveloped world as a laboratory, for that is imperialism in the true sense of the word.

REFERENCES

Angell, Marcia. "Ethical imperialism? Ethics in international collaborative clinical research." *New England Journal of Medicine* 319 (1988): 1081-1083.

________. "The ethics of clinical research in the third world." *New England Journal of Medicine* 337 (1997): 847-9.

________. "Patients' preferences in randomized clinical trials." *New England Journal of Medicine* 310 (1984): 1385-87.

________. "Tuskegee Revisited." *Wall Street Journal*. October 28, 1997: A22.

Annas, G.J. and M.A. Grodin, eds. *The Nazi doctors and the Nuremberg code: Human Rights in Human Experimentation*. New York: Oxford University Press, 1992.

Brennan, T.A. "Proposed revisions to the Declaration of Helsinki – will they weaken the ethical principles underlying human research?" *New England Journal of Medicine* 341 (1999): 527-31.

Brown, JG. *Recruiting Human Subjects: Pressures in Industry-Sponsored Clinical Research.* Department of Human Services, Office of Inspector General. June 2000. OEI-01-97-00195.

Chang, L. "China's pool of patients lures start-ups doing clinical trials." *Wall Street Journal*. April 9, 2002: D12.

Connor, E.M., R.S. Sperling, R. Gelber et al. "Reduction of maternal-infant transmission of human immunodeficiency virus type 1 with zidovudine treatment." *New England Journal of Medicine* 331 (1994): 1173-1180.

Council for International Organizations of Medical Sciences. *International Ethical Guidelines for Biomedical Research Involving Human Subjects*. Geneva: Council for International Organizations of Medical Sciences, 1993.

Department of Health and Human Services. Protection of Human Subjects, 45 CFR Part 46. ("Common Rule").

Freedman, B. "Equipoise and the ethics of clinical research." *New England Journal of Medicine* 317 (1987): 141-5.

French, H.W. "AIDS research in Africa: juggling risks and hopes." *New York Times*. October 9, 1997: 1.

Glantz, L.H. and M.A. Grodin. "Ethics of placebo-controlled trials of zidovudine to prevent the perinatal transmission of HIV in the third world." Correspondence. *New England Journal of Medicine* 338 (1998): 836-841.

Lallemant, M., G. Jourdain, et al. "A trial of shortened zidovudine regimens to prevent mother-to-child transmission of human immunodeficiency virus type 1." *New England Journal of Medicine* 343 (2000): 982-991.

Levine, R.J. "The need to revise the Declaration of Helsinki." *New England Journal of Medicine* 341 (1999): 531-4.

Loff, B. and M. Heywood. "Patents on drugs: manufacturing scarcity or advancing health?" *Journal of Law and Ethics* 30 (2002): 621-31.

Lurie, P. and S.M. Wolfe. "Unethical trials of interventions to reduce perinatal transmission of the human immunodeficiency virus in developing countries." *New England Journal of Medicine* 337 (1997): 853-6.

Pomfret, J. and D. Nelson. "The Body Hunters." *Washington Post*. December 20, 2000: A01.

Varmus, H. and D. Satcher. "Ethical complexities of conducting research in developing countries." *New England Journal of Medicine* 337 (1997): 1003-5.

World Medical Association. "Declaration of Helsinki." *Journal of the American Medical Association* 284 (2000): 3043-45.

II.

MEDICAL ETHICS

RUTH MACKLIN

REPRODUCTIVE RIGHTS AND HEALTH IN THE DEVELOPING WORLD[1]

Women in all parts of the world face obstacles to their ability to exercise reproductive rights and maintain reproductive health. This is true in industrialized, democratic societies as well as in developing countries and those with oppressive political regimes. It is nevertheless true that the obstacles are much worse in those parts of the world in which women are systematically oppressed, have few civil rights, or are in such dire poverty that they are unable to afford preventive and therapeutic services that would otherwise be available to them. Some of the obstacles to women's reproductive rights and health are legal and political; some are consequences of deeply rooted customs or cultural norms, including religious teachings; and others are a combination of several of these factors. This article will review some of these challenges confronting women and families.[2]

In a majority of developing countries, legal barriers restrict women's access to safe abortions. Laws prohibiting abortion for reasons other than to save the life of the woman exist in many countries, but as is well known, such laws do not prevent women from having abortions. They typically prevent women from having *safe* abortions, resulting in significant morbidity and mortality. At least 78,000 women die each year from complications of unsafe abortion and hundreds of thousands of women suffer from long- or short-term disabilities. In low-income countries, about 200 women die each day as a result of unsafe abortions. Unsafe abortions are responsible for 13% of all maternal deaths globally. Each year, an estimated 20 million unsafe abortions are performed worldwide, 95% of which are performed in low-income countries.[3]

[1] Portions of this article are excerpted from *Against Relativism: Cultural Diversity and the Search for Ethical Universals in Medicine* by Ruth Macklin, copyright © 1999 by Oxford University Press, Inc. Used by permission of Oxford University Press, Inc.

[2] Some of the accounts described below were products of workshops and interviews in which I participated over a period of four-to-five years. In two successive projects on ethics and reproductive health supported by the Ford Foundation, I visited each of the following countries at least once, making more than one visit to several: the Philippines, Mexico, Chile, Argentina, Peru, Colombia, China, Nigeria, Egypt, India, Bangladesh, and Brazil. Meetings and interviews took place with physicians and other health personnel, governmental officials, women's health advocates, representatives of nongovernmental organizations, lawyers, social scientists, and religious leaders. All references to interviews, meetings and the remarks of interviewees and meeting participants are taken from interim or final reports submitted to the Ford Foundation. The locations and dates of the meetings or conversations are provided in notes.

[3] Center for Reproductive Law and Policy, "Safe Abortion."

A. W. Galston and C. Z. Peppard (eds.), Expanding Horizons in Bioethics, 87-101.

Take Chile, for example. Although typically referred to as a "developing" country, Chile is not desperately poor, it has a large middle class, and there is a very high literacy rate. Yet along with Peru, Chile has the highest rates of abortion of any country in Latin America. Each year, almost one woman in every 20 (age 15-49) has an induced abortion, and there are close to six abortions for every 10 births. So in Chile, as elsewhere, prohibitive laws clearly do not prevent abortions from occurring. Yet the sanctions for choosing to undergo an abortion are severe for those women who are unfortunate enough to be identified. On one date in September 1997, 22 women were in prison in a facility in Puerto Montt for abortion-related offenses.[4] In some countries, legal restrictions also limit access to sterilization; and in some places, the policies of ministries of health or family planning clinics deny unmarried women access to contraceptive services. These legal and policy restrictions have been well documented, but they tell only part of the story. Non-legal barriers to gaining access to reproductive health services can be just as formidable, and often remain in place even in the absence or removal of legal restrictions.

REPRODUCTIVE RIGHTS

What are reproductive rights? And how can we justify claims to certain reproductive rights as deserving the status of "human rights?" International human rights treaties, especially the Convention on the Elimination of All Forms of Discrimination against Women (also known as the Women's Convention), impose at least the following obligations on governments: governments must ensure that women are free from all forms of discrimination, that they are granted the right to liberty and security, and have access to health care and the benefits of scientific progress. Taken together, these separate rights under international law converge into a composite right of reproductive self-determination, also known as the right to regulate one's fertility.[5] It is worth noting that the United States, which we typically consider to be one of the more "advanced" nations in respecting the rights of women, has not signed the Women's Convention. This means that our government does not recognize the rights embodied in that international convention as human rights for women in this country. The case of the United States will be addressed at the end of this chapter.

Women's reproductive rights were a key topic at two international conferences in the past decade: the 1994 United Nations Population Fund International Conference on Population and Development (ICPD) and the 1995 United Nations Fourth World Conference on Women (FWCW). The ICPD Programme of Action states the following:

> Reproductive rights embrace certain human rights that are already recognized in national laws, international laws and international human rights documents and other consensus documents. These rights rest on the recognition of the basic rights of all

[4] Center for Reproductive Law and Policy, "Women Behind Bars."
[5] Cook, 15-16.

> couples and individuals to decide freely and responsibly the number, spacing and timing of their children and to have the information and means to do so, and the right to attain the highest standard of sexual and reproductive health. They also include the right of all to make decisions concerning reproduction free of discrimination, coercion and violence.[6]

An example of a violation of the reproductive rights in the ICPD statement is a campaign launched by the Peruvian government to sterilize women in the late 1990s. A violation of this same right has taken the opposite form with regard to sterilization: laws that prohibit voluntary sterilization are still on the books in Argentina, a Latin American country where Roman Catholicism exerts a strong influence on the legislature. All such laws or governmental policies interfere with the reproductive freedom of individuals or couples, and are therefore in violation of the international covenants that include as human rights the right to liberty, the right to found a family, and the right to determine the number and spacing of one's children.

In some places, equally strong barriers to women seeking to exercise their reproductive rights come from within the family. The one-child policy in China, enforced more in urban areas than in the countryside, is a well-known example of a highly restrictive governmental policy. The official policy required that an intrauterine device be inserted in any woman who had more than one child; and women who had more than two children were to be sterilized. Yet even when women are willing to limit the number of children and accept the state-controlled limitations, their husbands may be the ones insisting on having a larger family. Gender preference is a further complication in China. And not only husbands but also the husbands' mothers effectively limit the choices women may make. The power of the mother-in-law in China came to light in one meeting I attended several years ago with representatives from the State Family Planning Commission. An example was provided by a physician who worked as an administrator in the Commission. She described the following case.[7]

A woman in a village had three children, all girls. She did not want more children, but her husband and mother-in-law did not want her to be sterilized, as they actively desired male children in the family. The ethical issue was presented as a dilemma for the physician: Should she sterilize the woman or follow the wishes of the patient's husband and mother in law? The physician tried to persuade the mother-in-law and husband to agree to the woman's sterilization, but that attempt failed. What would be the consequences for the woman if the physician counseled her to go ahead with the sterilization? Everyone at the Commission meeting concurred that if the woman were sterilized, the husband would divorce her in order to find a new wife who might bear him sons.

What would be the consequence for the woman following a divorce? She would be rejected from the family and lose whatever possessions she had, including custody of her own children, who would remain with their father and his family. Although this woman could theoretically remarry, she did not in fact want to divorce

[6] United Nations Population Fund.

[7] This meeting occurred on February 25, 1993 in Beijing.

the husband. She loved her family, and she especially did not want to lose custody of her daughters. This story is instructive, not only because of what it says about the subordination of women's wishes and claims within the traditional family in China, but also for the account of a divorced woman's loss of custody of her children and whatever material possessions she may have.

A somewhat similar situation can be found in traditional areas of Mexico, but stemming from quite different cultural factors. Reports from women activists in Mexico revealed that the dominant value of "machismo" creates in men the desire to prove their masculinity by having many children. Indeed, whether it is their wives or their mistresses who bear the children, macho credit goes to the men for fathering numerous offspring. A Mexican social anthropologist described the prevalent values of patriarchal Mexican culture as follows: "Until recent years, popular sayings included: 'To be a man is to be the father of more than four.'" A saying pertaining to women is considerably less charitable: "Women...were to be kept 'like shotguns, loaded and in the corner'. That is to say, pregnant and marginalized when it came to important matters."[8] Information provided by social scientists and women's health advocates revealed that in at least some cases, women seek to limit family size while their husbands, with strong support from the husbands' mothers, compel the wives to continue to have children.

As in China, the authority of the mother-in-law in Mexico is considerable, and goes well beyond the promotion of their sons' desire to have many children. A researcher conducting social science research on reproductive health in a very poor area described the need to understand the traditional beliefs and values of the people on whom the research is being carried out.[9] As an example, she noted that women in traditional settings have to ask permission from their mother-in-law to visit a physician. The mother-in-law may question that decision or refuse to grant permission. The woman might actually be expelled from her home if the mother-in-law finds out she has gone to a physician without her consent.

A quite different cultural factor explains the behavior of Nigerian women in choosing to have more children than they themselves desire. In many parts of Nigeria, polgyny (one man with multiple wives) is widespread, and birth control does not mean the same to women in those relationships as it does to women in a monogamous society. Some Nigerian women in the polygynous culture believe that if they practice birth control their husband will take another wife.[10] However, the women are mistaken in their belief that having more children will deter their husband from taking another, younger wife, since the husbands go ahead and marry again anyway—always a younger woman—regardless of how many children their earlier wives have borne.

One factor inhibiting women's opportunities for reproductive choice in some developing countries emanates from deeply entrenched views regarding "male superiority." Views about the natural inferiority of women remain widespread in the

[8] Elu, 59.

[9] The meeting took place on February 4, 1993.

[10] This was stated in a meeting that took place in Ibadan, Nigeria, with the Social Science and Reproductive Health Network on October 19, 1993.

Seychelles, to take only one example. Following a presentation on reproductive rights at a meeting in Jamaica of the Commonwealth Medical Association, a young physician from the Seychelles voiced vigorous objection to the very concept of reproductive rights for women. He had received his medical education in Czechoslovakia, and he commented on the disintegration of society that had resulted from granting women equal rights in that country. He claimed that encouraging women to work outside the home and to receive an education equivalent to that of men has led to a high divorce rate—he said that about 50% of the marriages end in divorce in Czechoslovakia. This young physician characterized the high divorce rate as the disintegration of the family, and hence, the society as a whole. Therefore, he argued, granting women reproductive rights, among other rights, is a sure way to undermine traditional cultural values and eventually, the entire culture.

This man buttressed his argument with a number of highly questionable and unsubstantiated claims about the "natural order" of things: "Men are naturally superior to women;" "men are stronger, more aggressive, more equipped to lead;" "women are naturally weaker, more subservient, naturally subordinate." He was convinced that the natural order of things—as he perceived it—made it perfectly reasonable to fashion societal norms in a way that reinforces this nature. He further proposed that women want men to be superior, to lead them, and to take responsibility on all levels. If "women want it that way," he argued, ostensibly women cannot therefore be made vulnerable or wronged by such arrangements.

DEMOGRAPHIC GOALS AND REPRODUCTIVE CHOICE

So far, I have described obstacles to women's reproductive choices that stem from governmental laws and policies, as well as cultural factors that continue to hold sway in traditional societies. But there have been international forces at play, as well. For about three decades—since the mid-1960s—an ongoing conflict has persisted between two prominent international movements. The first has been the movement to control the rates of increase of populations, especially in developing countries where birthrates have tended to be much higher than in industrialized countries. The second movement has been the UN-sponsored World Population Plan of Action, which since 1974 has included the statement (quoted above) from the 1994 world population conference: "All couples and individuals have the basic right to decide freely and responsibly the number and spacing of their children and to have the information, education and means to do so." Yet at the same time this reproductive right was being enunciated in the United Nations population conferences held every ten years, demographic goals were imposed in many countries in an effort to reduce the rate of population growth with the aim of reaching targets. Governments imposed these goals on family planning programs, which in turn placed restrictions on physicians who worked in the public sector.

Two Mexican doctors described poignantly and in detail the ethical dilemmas they faced in their work in the public sector.[11] Beginning with their training in

[11] This meeting of health sector personnel took place on February 2, 1993.

obstetrics and gynecology, physicians were instructed to insert an intra-uterine device (IUD)—a plastic or metallic ring, loop or spiral that is implanted into a woman's uterus as a method of birth control—in every woman who had delivered three children. Since many women might refuse the IUD if asked, a large number of physicians inserted the IUDs without informing women. Obviously, this behavior not only violated the women's right to refuse an unwanted intervention; it also violated their right to be informed about what physicians were doing to them. Two doctors reported their experience as members of a group of physicians who were told to insert IUDs in women or else they would be fired from the hospital. Some of these doctors "complied" by placing the IUDs in the women's vaginas rather than fully inserting the devices. The IUDs fell right out when the women stood up.

The two physicians who told this story left no doubt that they judged the mandated IUD insertion to be ethically wrong. Yet they complained that those in authority did not see the requirement as posing an ethical problem. The authorities simply imposed a quota, and the physicians felt they must fulfil it. People in several countries I visited said that this top-down pressure has been a consequence of international loans from agencies such as the World Bank and the International Monetary Fund. Targets for reducing population growth were set through international agreements, with penalties for countries that failed to meet their pre-determined quotas. The financial pressure on governmental leaders in developing countries like Mexico made limitation of population an overarching public policy objective in the past three decades.

The good news is that the international movement to control population has gradually been replaced by a more enlightened approach that focuses on reproductive health. Yet even acknowledging the coercive nature of population control programs of past decades, one commentator wrote recently that "between the late 1960s and the early 1990s national family planning programs became one of the major public health successes of the 20th century."[12] These programs have without doubt benefited women in many countries by meeting their desires for smaller families and the means to achieve that goal.

RELIGIOUS BARRIERS TO REPRODUCTIVE RIGHTS

In many parts of the world, religious barriers to women's access to effective contraception illustrates the violation of a reproductive right in an opposite manner from that of coercing women into accepting contraceptives such as the intrauterine device. The IUD is not widely accepted in the Philippines because most people, including politicians and even physicians, consider the device to be an abortifacient. This is a widespread belief despite published research demonstrating that the IUD prevents fertilization rather than implantation of the fertilized ovum.[13] The Chairman of the Department of Obstetrics and Gynecology at the University of the Philippines wondered whether his departmental program could require medical

[12] Rosenfield, 1838-1840.

[13] This research was reported in "Research Roundup," which reported on the article by Chi.

students and trainees to learn how to insert IUDs. In a country inhabited by many deeply religious Roman Catholics, a student or an ob/gyn resident might object to artificial contraception, and hence to learning how to insert IUDs in patients. Even among those who are not opposed to artificial contraception, medical students and trainees who believe—contrary to medical science—that an IUD acts as an abortifacient would refuse to learn the procedure on that ground. This illustrates the sort of dilemma that can arise when the requirements of good professional training appear to conflict with a religious prohibition.

In Chile also, restricted access to contraceptives is a result of the strong influence and power of the Catholic Church. Yet Chile is a country in which population growth had long been stabilized, thus precluding the need for state-imposed limitations on access. The role of the church came up in a meeting I attended with members of a coordinating body for twenty-odd women's organizations. It was noted there that the position of the church regarding the use of contraception is: "you are going to kill a son." One participant said that Chile received millions of dollars from international donor agencies for IUDs, but the devices were never distributed. This failure to distribute the IUDs was attributed to the action of the church as well as to the government in power, which goes along with the pro-natalist position of the church.

Given this background, it is perhaps surprising that Chile has one of the lowest fecundity figures in Latin America. One person at the meeting of women's groups that I attended said that the government is hypocritical in this regard: Chile publicizes its low fecundity rate with some pride. Yet the government does not provide contraceptives and there are thousands of illegal abortions performed annually. The hypocrisy lies in the fact that the government is pro-natalist in its policies and their implementation, and yet at the same time it boasts of the country's low fertility rate.

In another meeting, organized by the Chilean office of the United Nations Population Fund,[14] one person noted that groups of midwives refuse to give contraceptives to women as part of family planning services. The midwives cite "conscience," stemming from their Catholic belief system, as the reason why they should not be required to provide the contraceptives. Another speaker cited the argument the Vatican introduced into the International Conference on Population and Development in Cairo in 1994 and the Fourth World Conference on Women in Beijing in 1995. The Vatican's strategy was to argue that among poor women in developing countries, informed consent is impossible so women were being exploited in being offered contraceptives. How, the Vatican asked, can there be "informed consent" from women in countries where the power and authority of physicians eclipses the decision-making ability of women in poverty? Wouldn't the women feel compelled to accept whatever reproductive options the physicians offer? The church argues that when doctors ask poor women to do something, the physician is seen as "close to God," and therefore a woman does not have a genuine opportunity to refuse what the physician proposes. Although this description of the power relationship is accurate, the argument is specious. It essentially says that in

[14] This meeting took place in Santiago on October 2, 1995.

such situations, where power relationships are unequal, doctors should not even offer potentially beneficial treatments or procedures because the women could not freely refuse. But what is the alternative? The Vatican does not offer a positive approach, such as education; instead, they take the negative approach of refusing to offer the treatment or procedure at all, thereby reducing women's options, however limited they may be. The Vatican's unhelpful and circular argument effectively bars women from access to contraceptives.

The influence of the Vatican's pronouncements and mode of argument is palpable in countries with a strong Roman Catholic tradition, like Chile. In a meeting at the University of Chile's Centro Interdisciplinario de Bioética[15] (CINBIO, an Interdisciplinary Center of Bioethics), as soon as the phrase "reproductive rights" was uttered, a woman spoke up using a tactic I had not previously come across in Latin American countries. She said that the phrase "reproductive rights" is a code expression to allow women to "kill their unborn babies." My first inclination was to distinguish patiently between babies and fetuses, but the others in this group at the University of Chile were sophisticated intellectuals not given to invoking the slogans propagated by the Vatican or other right-to-life political groups. I deflected the question politely and went on to discuss the other issues that were on the table. Shortly thereafter, the woman who made the intervention departed.

And how different is it here in the United States? For years, most health insurance did not cover contraceptives, including expensive pills and other methods used by many women. It was not until the publicity surrounding Viagra, and the decision of health insurers to pay for that treatment for men with sexual dysfunction that a hue and cry arose, resulting in an about-face by some health insurers regarding their coverage of contraceptives. Yet as recently as May 2001, when a decision to provide medical insurance for contraceptives was pending in the New York state legislature, New York's newly appointed Cardinal Egan went to the state capital to lobby against the provision. The New York state Assembly had passed a bill that would mandate prescription contraception coverage without an exemption for religious organizations, whereas the state Senate had passed a bill that included the religious exemption. The Cardinal and his fellow Catholic lobbyists were seeking to influence the outcome of this secular legislation in a pluralistic, democratic state.

Religious organizations have sought to exert their influence on the use of contraceptives not only to prevent unwanted pregnancy, but also to prevent transmission of the AIDS virus. In Zimbabwe, Christian churches grouped together to oppose the AIDS policy of the ministry of health, which recommended the use of condoms to slow the spread of HIV/AIDS.[16] The chairman of a committee in this consortium of churches stated that condoms are not the solution to the spread of the disease, and accused the Zimbabwe government with disseminating "half-baked truths about the safety of condoms."[17] As recently as January 2001, the Catholic Church hierarchies in Zambia and Brazil have opposed AIDS prevention campaigns

[15] This meeting occurred on October 10, 1995.
[16] *In Catholic Circles* 1997, 3-4.
[17] Ibid.

that promote the use of condoms. One Brazilian bishop said that handing out free condoms did not arrest the spread of AIDS, but rather propagated it. Bishops in Guatemala opposed a plan to reduce infant and maternal mortality and combat the spread of HIV/AIDS. The president of the bishops' conference said that the plan would promote promiscuity, ignore paternal responsibility, and threaten the primacy of God and the family.[18]

SPOUSAL AUTHORIZATION: A NON-LEGAL BARRIER TO ACCESS

Other long-standing customs that restrict women's reproductive rights remain prevalent in developing countries. In these places, many people—including physicians—believe that existing laws prohibit providing contraceptives or performing sterilizations on women without the permission of their husbands. This barrier exists for women in the Philippines who seek sterilization when their family size is complete. An attorney from Manila said that the family code in the Philippines does not require authorization of a spouse for his or her partner's sterilization, yet the spirit of the law is not borne out in practice. Family planning clinics do, as a matter of fact, always require spousal permission.[19] Further, it is typically the husband's authorization that is sought for the wife's sterilization, as very few vasectomies have been performed in the Philippines.

In Nigeria, as well, the husband's signature is always obtained for a tubal ligation, even though the husband's permission is not a legal requirement. Participants at a workshop in Nigeria said that some women choose to come for the procedure without their husbands.[20] These participants, and the larger assembly, seemed to agree that a woman does have the right to obtain a tubal ligation without her husband's permission. But when the question arose whether the husband may go to court if the woman has a tubal ligation without his consent, the reply was that according to the official law in the villages, a man may seek to divorce a woman who does not bear him a child. Thus, while in theory a woman may exercise her own reproductive rights, in actuality the cultural norms erect strong barriers to women's reproductive choice. This is yet another example of how powerful cultural traditions persist and conflict with emerging health and reproductive rights—even when members of that same culture are in agreement over basic rights. Particularly when the members seeking such treatments are primarily women, their subordinate social status tends to render them powerless to change cultural customs.

A lawyer in Nigeria reported that physicians honor spousal authorization even in making contraceptives available to women.[21] A woman's husband must come to the hospital or clinic to provide authorization for the wife's IUD (or other birth control method). If the husband has refused to come to the hospital for that purpose, the physician will not provide the woman with the contraceptive. This male control

[18] *In Catholic Circles* 2001, 3-4.

[19] This interview took place on September 9, 1992.

[20] This workshop was held on October 20, 1993.

[21] The conversation took place on October 15, 1993.

over reproduction is evident in traditional Nigerian culture, in which the woman is not viewed as the person who has any claim over the children. Rather, children "belong" to the husband. The language of ownership is the basis for the requirement of spousal authorization for a reversible method of birth control, such as contraceptives.

Such practices are often defended in the spirit of respect for tradition and the need to demonstrate cultural sensitivity. While I would agree that it is always appropriate to be sensitive to traditional customs and culture, an ethical obligation exists to draw a line when respect for cultural tradition interferes with the reproductive rights of an individual. To defend a violation of women's reproductive freedom by appealing to culture or tradition is not to provide an ethical justification. It is simply to argue that a cultural tradition should remain a cultural tradition. That is no ethical argument, but is rather a defense of the *status quo*, and it effectively undermines the primacy of reproductive—and human—rights.

JUSTICE IN ACCESS TO REPRODUCTIVE HEALTH SERVICES

Underlying virtually all of the issues explored in this brief overview are questions of justice. Given the great disparity between what is available to wealthy people and to poor people, in industrialized as well as in developing countries, questions of justice always lurk in the background. For decades, people have argued that it is ethically justifiable to approve some restrictions on reproductive rights in developing countries threatened by a high rate of population growth. A reply to that argument points out that such restrictions are invariably applied only to poor women in those countries. Placing restrictions on reproductive rights frequently results in injustice, given the ability of financially well off members of any society to obtain medical services or avoid governmental restrictions.

A barrier to access to an acceptable quality of reproductive health services exists in countries with a poor health infrastructure. India is one such example. A health care worker from Parivar Seva Sanstha (PSS), a social service organization, said that 50% of couples in India never discuss contraception, sex, or family planning with one another. Typically, a woman seeks such information and services on her own. So the most common pattern is that following her marriage the woman does not use contraceptives, she gets pregnant, and then has an abortion. With a lack of awareness of contraception, abortion becomes a primary method of family planning. Abortions are readily available. 90% of contraceptive users are women who obtain contraceptives immediately after having an abortion. By law, untrained providers may not legally do abortions in India. Yet for every legal abortion, there are at least 9 more illegal and usually unsafe abortions. 11% of maternal mortality is due to septic abortions. Extra training is needed even for obstetrician-gynecologists who perform abortions, since in general, even physicians do not have much practice and training. Physicians are trained only in hospitals, never in small rural settings. In those places, therefore, women have no access to safe, legal abortions since there are only primary health centers with no personnel trained to do abortions. Nurse-midwives and traditional birth attendants perform abortions, many with inadequate skills. Thus in India, a country in which abortion is legally available, there is still a high morbidity and mortality from abortions. In contrast to countries in which restrictive abortion laws were liberalized as a consequence of efforts by the women's movement, in India it was the government that promoted decriminalization.[22] Though this decision has potential for affirming the reproductive rights of women, some commentators have suspected that the Medical Termination of Pregnancy Act of 1972 was intended as a tool for population control. A report from 1997 documents a number of ways in which the provisions of the law are at cross-purposes with the needs and aspirations of women. The motives and methods of social change are important to consider in the implementation of reproductive rights.

The situation regarding equitable access to healthcare in China has become steadily worse. In response to the question, "Which ethical issues are paramount in China?" a Beijing physician replied that the most important ethical issue in China is justice in the allocation of resources.[23] She said that less than 10% of the population can get medical care free of charge. However, 50% of those living in cities have free care. This is a maldistribution, since 80% of the entire population of China resides in the countryside. Anyone who is a government employee gets free medical care and everyone else is required to pay. Further, the physician said that the costs of medical care vary. Years ago, medical care was not very expensive, amounting only to about 10% of one's income. However, costs are much higher today, resulting in part from excessive use of medical technology due to economic interests. The physician gave the example of CT scans: many Western high-technology modalities are now available in China and, as in the United States, hospitals earn more money the more tests they do, so physicians are encouraged to do more high-technology procedures—even when not medically necessary.

Another example is the increasing cost of drugs. Government employees receive subsidized drugs manufactured in China. Traditional Chinese medicines are also used, and these are cheaper than pharmaceuticals. However, in rural areas where the farmers are still very poor (especially in the northeast and northwest), no one can afford drugs. The physician described this entire situation as one of maldistribution of medical resources and unequal access to the health system. She acknowledged that such inequalities have always existed in China, and though the minister of health would like to change the situation, it is difficult to bring about change.

As in other developing countries, physicians who work in the countryside have limited skills. They have received a different education from physicians who work in the city. During the Cultural Revolution, barefoot doctors received training in hospitals and were popular in the countryside. In addition, famous physicians from the cities were forced to go to the countryside to work. Today, there is low satisfaction and low income in medicine, so young people choose a career in business or another pursuit where they can earn more money than they could as a village doctor. As a result, many people who live in the countryside go directly to city hospitals when they get sick, and hospitals are eager to treat these patients in

[22] Gupte, Bandewar, and Pisal.
[23] The interview was on February 26, 1993.

order to earn more money. This results in even more crowding in outpatient clinics in city hospitals.

In addition, the change from a centrally planned economy to a free market economy has resulted in reduced welfare available to the Chinese people, as the government now controls less of the nation's capital for redistribution. Moreover, the government has expenses outside the health care system, such as building roads and attending to the infrastructure. The chief priority in China today is economic growth, and a commitment to health care has suffered as a result. In China, as in other developing countries, the gap between rich and poor is widening rather than narrowing, thereby worsening existing injustices. Although reproductive health is only one area in which barriers to access to care remain formidable, it is an area that affects large numbers of people—women disproportionately more than men. As this article has sought to demonstrate, the non-legal barriers to access are more difficult to surmount than are restrictive or oppressive laws.

Lest we who live in a wealthy, industrialized country be too complacent, we need to recognize that inequalities in women's access to health care and reproductive health services are not confined to developing countries. An article entitled "Human Rights Is a U.S. Problem, Too"[24] documents the limited access to medical care of HIV-infected women in the United States, most of whom are poor. 81% of women recently diagnosed with AIDS in the U.S. are black or Hispanic. The article faults the public health prevention messages from the government as "punitive and rigid" and the government's response generally as paternalistic and condescending. As an example, the article cites the U.S. public health message insisting that women convince men to wear condoms 100% of the time or refuse sex. Yet, the author notes, most women at greatest risk of HIV infection are unable to accomplish these preventive steps. The female condom, whose potential to reduce the risk of HIV infection has been noted by the World Health Organization as well as by many developing countries, has received little backing either financially or ideologically by federal public health agencies in the U.S.

What is our government doing to ensure that poor women in this country and abroad have access to reproductive health services? The answer is "not much." First we have the long-standing ban instituted by the U.S. Congress on the use of federal funds for abortion for poor women who lack private insurance. There still exists the "negative right" to abortion guaranteed by the Supreme Court's decision in Roe v. Wade—the right of a woman not to suffer state interference if she chooses to have an abortion within the legally imposed time limits. But there is no corresponding "positive right" that would require the state to provide or pay for such services for the segment of the population that is uninsured or receives federally funded insurance for other medical procedures.

The current administration has reinstated a powerful move against poor women all over the world. The "Global Gag Rule" was instantiated in the Reagan era and has been resurrected by the George W. Bush administration. The Global Gag Rule prohibits foreign organizations that receive U.S. funds from using their own, non-U.S. funds to provide legal abortion services, lobby for abortion law reform, or

[24] Gollub.

provide counseling or referrals on abortion. Most significantly, the Global Gag Rule turns back the clock on the recent worldwide trend toward liberalizing abortion laws. Experts agree that the effect will be exactly the opposite of what is intended by the law. Rather than decreasing the number of abortions that occur in these developing countries, the gag rule could very well lead to women becoming pregnant and seeking unsafe, illegal abortions because of limited access to effective counseling and provision of contraceptives. Spokespersons for women's reproductive rights and health in developing countries are outraged by this action on the part of a democratic government. One woman who directs a women's health advocacy group in Peru stated:

> The relationship of partnership with the United States has been compromised. The Global Gag Rule changed the partnership to "father-ship." It implies that except for institutions in the U.S., all the rest of the institutions are like children, so the U.S. can tell us what we can do with our money.... My country has the third highest maternal mortality rate in the region. I cannot even discuss this with legislators in my country due to the Global Gag Rule. And of course I am also unable even to stand here in your country—where you so value free speech—and discuss openly the reasons that high maternal mortality and unsafe abortion rates continue to impact so many Peruvian women. I do not want to endanger funding for the thousands of women our project is serving. We are against the Global Gag Rule because it prohibits the possibility of talking about a public health problem, even if the result of the discussion and advocacy is not to legalize abortion. As the U.S. should know, democracy is nourished and strengthened with free speech.[25]

Beginning with the 1994 International Conference on Population and Development in Cairo, there has been an escalating worldwide commitment to promoting and achieving reproductive rights for women. This has included provision of safe and accessible abortion services where abortion is legal, and the reform of restrictive abortion laws in other places. Despite the great strides that have been made in affording women in developing countries access to family planning services, disparities between rich and poor remain. Barriers to women's right to control their own reproduction generally include poverty, lack of education, and the low social status accorded to women—all perpetuating the injustices that have obtained for centuries. Nearly 600,000 women die each year in poor countries either because they do not have access to safe abortions (even in countries where abortion is legal, such as India), or because of substandard skill of birth attendants and insufficient childbirth facilities.[26]

CONCLUSION

Let me end on a more positive note. Three recent movements for change show promise for improving women's reproductive health and strengthening women's reproductive rights. The first movement is an ideological shift on the world stage from a demographic emphasis on population control to the more positive goal of

[25] Statement by Susana Galdos Silva, Co-Founder of Movimiento Manuela Ramos, and Executive Director of Repro Salud.

[26] Germain.

promoting reproductive health. The second is an effort within countries to integrate family planning services into more comprehensive programs for women's health, not limited to reproductive health. And the third is the increasing recognition of women's rights and gains in women's empowerment that are occurring slowly yet perceptibly in many parts of the world. When these three movements coalesce and gain strength and monetary support, we can begin to realize our hopes for great strides in women's reproductive health and rights.

REFERENCES

Center for Reproductive Law and Policy. "Safe Abortion: A Public Health Imperative" (March 2000). http://www.crlp.org/pub_fac_atksafe.html.

________. "Women Behind Bars: Chile's Abortion Laws." http:// www.crlp.org/pub_bo_chilesum.

Chi, I-Cheng, "What We Have Learnt From Recent IUD Studies: A Researcher's Perspective." *Contraception* 48 (1994): 81-107.

Cook, Rebecca. "Putting the 'Universal' into Human Rights." *Populi* 21, no. 7 (1994): 15-16.

Elu, Maria del Carmen. "Abortion yes, Abortion no, in Mexico." *Reproductive Health Matters* 1 (1993): 58-66.

Germain, Adrienne. "Population and Reproductive Health: Where Do We Go Next?" *American Journal of Public Health* 90, no. 12 (2000): 1845-1847.

Gollub, Erica L. "Human Rights Is a U.S. Problem, Too: The Case of Women and HIV." *American Journal of Public Health* 89, no. 10 (1999): 1479-1482.

Gupte, Manisha, Sunita Bandewar, and Hemlata Pisal, "Abortion Needs of Women in India: A Case Study of Rural Maharashtra." *Reproductive Health Matters* 9 (1997): 77-86.

In Catholic Circles: An International News Roundup. 2 (6) 1997: 3-4.

________. January-February 2001: 3-4.

Macklin, Ruth. *Against Relativism: Cultural Diversity and the Search for Ethical Universals in Medicine.* New York: Oxford University Press, 1999.

"Research Roundup." *Reproductive Health Matters* 3 (1994): 115.

Rosenfield, Allan G. "After Cairo: Women's Reproductive and Sexual Health, Rights and Empowerment." *American Journal of Public Health* 90, no. 12 2000: 1838-1840.

Silva, Susana Galdos. Official Statement. February 14, 2001. http://www.crlp.org/pr_01_0214ggrsilva.html.

United Nations Population Fund. *Summary of International Conference on Population and Development Programme of Action.* Chapter VII, "Reproductive Rights and Reproductive Health." http://www.unfpa.org/icpd/summary.htm#chapter7.

KATHY HUDSON, SUSANNAH BARUCH & GAIL JAVITT

GENETIC TESTING OF HUMAN EMBRYOS

Ethical Challenges and Policy Choices[1]

INTRODUCTION

Today, there are some one million people for whom the journey toward personhood began when a fertility specialist, peering through a microscope, carefully added sperm to egg in a glass petri dish—a process known as *in vitro* fertilization (IVF). There have also been dramatic advances in our scientific understanding of the human genome during this time period, which has led to the ability to test for genetic alterations associated with diseases and other inherited characteristics. Currently there are tests for over 1000 genetic diseases available or under development, and the number is steadily growing.[2]

The independent fields of IVF and genetic testing each present a host of issues that are technically, legally and ethically complicated. But now, the worlds of genetic testing and assisted reproduction have converged, most notably in preimplantation genetic diagnosis (PGD)—technology that allows parents to choose which embryos to implant in a woman's womb based on genetic test results. The arrival of PGD has engendered a host of new scientific, social, ethical and political quandaries. Many people have begun to consider not just the implications of this new genetic diagnostic tool but whether core ethical and practical concerns surrounding IVF are really all that settled.

Adding genetic testing to the IVF process means that medical providers and scientists can now be deeply involved in the molecular mechanics of the most profound and mysterious of human activities: creating life. This intercession of technology into human reproduction evokes a range of responses. For some, it is a deeply offensive act in which science literally subsumes the role of God. For others, it allows science to alleviate the anguish of genetic disease and infertility. These dueling opinions reveal both PGD's potential benefits and its possible risks. As with much modern scientific research, there is a basic tension between concerns about the adverse consequences of unregulated research and fears that we may fail to develop important technologies if we apply too much restraint. Regardless of how one feels about it, PGD is a powerful tool, for it allows parents to identify and select the

[1] This chapter was written with the generous support of the Pew Charitable Trusts. The opinions expressed are those of the authors.

[2] http://www.genetests.org/.

A. W. Galston and C. Z. Peppard (eds.), Expanding Horizons in Bioethics, 103-122.

genetic characteristics of their children. As a society, we must ask whether and under what conditions PGD should be used.

In this chapter, we provide an overview of PGD and an analysis of the ethical, legal and social issues that surround it. We then present an array of policy options that could be employed to govern PGD's development and use. This essay surveys the situation of PGD in order to encourage discussion and facilitate policymaking; it does not endorse one course of action over another. The challenge with new biomedical technology is for the public and policymakers to keep informed despite the rapid pace of change. We believe that new genetic technologies that alter how we have babies, and indeed what babies we have, are so important that the public and its representatives must be engaged in the discussion and formulation of policies.

I. WHAT IS PREIMPLANTATION GENETIC DIAGNOSIS?

PGD is a process in which embryos developed outside of the womb are tested for particular genetic characteristics, usually genetic abnormalities that cause serious disease, before being transferred to a woman's uterus.[3] It is always performed within the context of IVF, for reasons that will be developed below.

PGD derives from recent advances in both reproductive medicine and genetics. In 1978, scientists achieved the first viable human pregnancy from an egg fertilized outside the womb. Around this time, geneticists' understanding of the basis of inherited disorders increased. Genetic researchers developed a number of tests to detect specific disorders. Eventually clinicians were able to apply these tests to a small amount of genetic material taken from an egg or embryo. The use of these tests in a human embryo *(in vitro)* or fetus (*in utero*) enabled prenatal detection of genetic abnormalities.

PGD is a multi-step process involving egg extraction, in vitro fertilization, cell biopsy, genetic analysis and embryo transfer. Eggs removed from the mother (after she has been given drugs to stimulate egg production) are fertilized in the laboratory. Once the egg is fertilized *in vitro*, it develops into an embryo. Most commonly, genetic tests are performed on one or two cells taken from an embryo two to four days after fertilization.[4] The sample may be analyzed two ways: by chromosomal analysis to assess the number or structure of chromosomes present in the cells; or by DNA analysis to detect specific gene mutations. Regardless of the methods, the results of preimplantation genetic diagnosis are used to inform the selection of embryos for transfer to a woman's uterus.

PGD enables two types of reproductive decisions. First, it permits doctors and prospective parents to select embryos for implantation that do not have a genetic abnormality associated with a specific disease, such as cystic fibrosis. Second, it

[3] Munne and Wells, 239.

[4] Handyside and Delhanty, 271. Alternatively, genetic tests can be performed on polar body cells, which are cast off by the egg as it matures and is fertilized. Verlinsky and Kuliev, "Current Status," 13.

enables doctors and parents to select embryos that possess a desired genetic trait, such as a tissue type that matches that of an ailing sibling.

Since PGD was first made available to facilitate embryo selection, more than 1,000 babies have been born worldwide following a preimplantation genetic test.[5] PGD was initially developed to detect serious single gene disorders,[6] not to alleviate infertility. More recently, however, PGD has been used as an adjunct to standard IVF to detect abnormalities in chromosome number, called aneuploidy, that arise during egg or embryo development. Some IVF providers recommend PGD for patients over 35 or those with repeated IVF failure.[7] Since over 1% of all U.S. newborns are IVF babies, and since more than half of IVF patients are of advanced reproductive age,[8] aneuploidy screening likely accounts for the biggest growth area in the use of PGD. There is no required tracking of the number of PGD procedures performed or of the purpose for which they are performed, however, so a definitive breakdown of how many PGD procedures are performed to detect single gene mutations versus aneuploidy is not available.

Virtually any of the hundreds of genetic tests now commercially available (and the many more in development) could be used in PGD. The more tests available, the more options there are for embryonic selection. Possible (but controversial) applications of PGD include its use to select an embryo that is an immunological match for a sick sibling;[9] to select the sex of an embryo in the absence of a sex-linked disease risk;[10] to test embryos for gene mutations associated with diseases such as early-onset Alzheimer disease[11] or Huntington disease[12] that do not appear until later in life; or to test for mutations that indicate a heightened but uncertain risk of developing a particular disease such as cancer.[13]

There are inherent limits to the use of PGD to avoid disease or select for certain traits. First, not all diseases have a clearly diagnosable genetic component. Many diseases and traits are the result of a complex interaction between multiple genetic and environmental factors. Second, if PGD is used to detect genes that indicate a heightened risk for a particular disease, such as hereditary breast cancer, the fact that such a gene is detected in the embryo does not mean that a person who developed from that embryo would definitely develop the disease. Finally, it is important to remember that PGD cannot create new genetic characteristics that neither parent has. PGD can allow parents to select only among the genetic combinations present in the embryos they have produced.

[5] Kuliev and Verlinsky, “Thirteen Years’ Experience,” 229.

[6] The first PGD cases were performed to determine embryo sex, in order to avoid X-linked disease. (Munne and Wells, 239). Other early uses included detection of genes causing cystic fibrosis, Tay Sachs disease, and Lesch-Nyhan syndrome (Verlinsky et al., “Preconception,” 103-110; Delhanty, 1217-1227).

[7] Ferraretti et al., 694-699; Munne, S70-S76.

[8] Kuliev and Verlinsky, “The Role,” 233.

[9] Verlinsky et al., “Preimplantation Diagnosis for Fanconi Anemia,” 3130-3133.

[10] Robertson, “Extending Preimplantation Genetic Diagnosis: The Ethical Debate,” 468-470.

[11] Verlinsky et al., “Preimplantation Diagnosis for Early-Onset Alzheimer Disease,” 1018-1021.

[12] Sermon et al., 591-598.

[13] Rechitsky et al.. 148-155.

II. ETHICAL ISSUES

PGD raises important concerns related to whether and when it should be used, its safety and effectiveness, costs and access, and what it would mean to live in a society where one's genetics become more a matter of choice than chance. Current oversight of PGD's safety, accuracy and effectiveness is extremely limited, as is the practice of IVF in general. Limits based on ethical, moral or societal concerns are almost nonexistent. The question is whether additional oversight is needed and if so, what issues it should address and how it should be structured. These are complicated dilemmas about which there has been little discussion or opportunity to form agreement.

Since PGD requires IVF, it is mainly used today by a relatively small number of women who are willing to undergo IVF to avoid a known serious or fatal genetic condition or who are unable to get pregnant without IVF because of infertility problems. For people who already turn to IVF to treat infertility, PGD may become a more common tool to screen for genetic variants ranging from the serious to what some might consider trivial. For the moment, one would expect very few prospective parents who do not need IVF to get pregnant and who do not seek to avoid a serious or fatal known genetic condition to utilize PGD. Nonetheless, as the number of genetic tests that can be employed successfully in PGD increases and IVF techniques improve, it is possible that more prospective parents may consider using IVF and PGD to "choose" their embryo.

Hence, with the advent of new reproductive biotechnologies—of which PGD is exemplary—there is likely to be growing public interest in developing policies that address ethical, technical and social concerns raised by the genetic testing of embryos. Below, we summarize the range of concerns about PGD and outline the possible policy alternatives that could guide the future development and use of this technology.

Is PGD "ethical"?

The ethical and moral ramifications of PGD have attracted significant attention. Some categorically oppose PGD's use under any circumstances,[14] while others have focused on the circumstances under which it may and may not be acceptable.[15]

As noted previously, PGD enables the selection of one or more embryos over others on the basis of genetic makeup. Embryos deemed genetically undesirable will likely be destroyed if they are not frozen indefinitely. Thus, PGD and the underlying IVF process involve the creation and, frequently, the destruction of human embryos. PGD is therefore objectionable to people who believe that a unique human being, deserving of the protections of a born person, is formed at the moment a sperm fertilizes an egg. From this perspective, no use of PGD is truly

[14] Catholic Health Association, 16.

[15] Robertson, "Extending Preimplantation Genetic Diagnosis: Medical and Non-Medical Uses," 213-16; Adams, 489-94; Botkin, 17-28.

"therapeutic," because the testing does not treat the condition it detects. Rather, it diagnoses a "patient" with the sole purpose of telling parents which "patient" to discard. People who hold that the sanctity of human life extends to conception therefore tend to oppose PGD.[16]

There are some people who do not hold a firm position on the moral status of the early human embryo but who nonetheless oppose PGD because they view it as unnatural or as violating the ways of nature. Still others argue that we should be wary of PGD (even if it is not inherently wrong or offensive) because it places society atop a slippery slope that will lead to genetic enhancement and human control of reproduction.[17]

The range of possible oversight proposals parallels the range of ethical opinions on PGD. For those who categorically oppose PGD, a permanent ban may be the only satisfying approach. A federal or state ban on PGD would have the benefit of creating a clear rule for prospective parents, health care providers and society. However, an outright ban would impose a single moral or ethical viewpoint, raise Constitutional concerns, and could be difficult to enact and enforce.

For those who are concerned about the societal implications of PGD for certain uses, a temporary ban—to be revisited after society has more carefully considered the implications of this technology—might alleviate these concerns. Other possibilities include greater federal, state, and/or professional oversight of PGD, both to ensure its safety and effectiveness and to limit its uses to those deemed acceptable to society.

Questions About Specific Uses of PGD

PGD is now used primarily to increase the chance of having a child free of a specific serious disease. But there are no legal limits on the use of genetic tests in PGD, and the technology can be applied to choose a child with certain traits, such as sex and HLA type.[18] Although some providers believe that certain uses of PGD are unethical and refuse to do PGD under certain circumstances (for example, to select an embryo of a particular sex for non-health-related reasons), others advertise these services and believe that parents should have the freedom to decide what uses of PGD are appropriate to their particular needs.

Some observers argue that parents have always tried to give their children every possible advantage, from vitamin supplements to private swimming lessons. PGD, they argue, should therefore be viewed as a technology that simply extends the

[16] It should be noted that these questions first emerged with the advent of IVF, though they have often been subsumed by the perceived good of enabling otherwise-infertile couples to bear children. In both IVF and PGD embryos will be discarded or frozen indefinitely if they are unused; in PGD, however, embryos are affirmatively rejected if they are deemed "unacceptable" because of the presence of a particular mutation or the absence of a particular desirable trait.

[17] The President's Council on Bioethics raises this concern in its book *Beyond Therapy,* in which the Council addresses PGD as part of a larger project on biotechnology (see especially pages 40-57).

[18] Human Leukocyte Antigens are proteins on the surface of white blood cells that play a role in immune response. The antigens are used to determine the suitability of a match between a donor and a recipient. Since antigens are inherited from parents, siblings are more likely to match than strangers.

boundaries of this natural tendency. Another view suggests that PGD is appropriate when used to avoid serious genetic disease, but is inappropriate when used to detect mild conditions or benign traits. On this logic, the use of PGD to avoid suffering outweighs the risks involved and concerns over the status of embryos. But the balance tips when PGD is applied to avoid a mild or treatable condition, or used to select embryos for particular traits deemed "desirable" but whose absence does not cause suffering.

The discussion of appropriate uses is made more complex because the lines between a serious health problem, a mild or treatable disease, and a trait unrelated to disease are diffuse and often semi-permeable. Two related questions emerge: Where is the line to be drawn between acceptable and unacceptable uses of PGD? And who should draw this line? As but one example, many people in the medical community would say that a genetic predisposition to hearing loss represents a serious medical condition. Yet many in the Deaf community consider deafness to be a culture, not a disability.[19] Should physicians, parents, or the government be the gatekeepers of "appropriate use" of technology? Whose values should have priority: the physician who sees deafness as a disability, the government who (in theory) sets the parameters for use of technology, or the parents who want to avoid, or in some case select for, a deaf child? It is clear that PGD raises the question of who decides what a "good life" is, and how far we should go in its pursuit.

Some people question the ethics of using PGD to screen embryos for diseases that will not affect a person until adulthood (such as Huntington disease). They reason that children born today with those mutations would enjoy several decades of normal health before any symptoms begin, during which time science may find a treatment or cure. The same question holds for genetic mutations associated with a heightened risk (as opposed to a certainty) for developing a particular disease. Should embryos be tested for a genetic mutation linked to an elevated risk (but not certainty) of developing hereditary breast cancer? Should an embryo with that mutation be summarily discarded?

Safety, Accuracy, and Effectiveness

For people who have decided to use PGD, the questions turn from broad ethical quandaries to more immediate issues such as safety and effectiveness. There are three chief concerns. Is this procedure safe for the mother and the resulting child? Does it accurately detect the genetic mutation of interest? And is it effective in producing a child free of that disease? Exploring these matters requires a consideration of the technical challenges and risks inherent in the genetic test itself and in the IVF procedure that it entails.

The data, however, are far from clear. There are incomplete and conflicting data concerning the risks IVF may present to mothers who undergo the procedure and the children conceived via this method. This situation makes it difficult to determine the extent to which adding PGD to the IVF process may introduce additional risks.

[19] Middleton et al., 1175.

In all IVF processes, there are risks associated with the hormones used to stimulate ovulation, and there is the risk the procedure could result in an ectopic pregnancy (in which the fetus develops in the fallopian tubes of the mother, and not in the uterus). Because more than one embryo is usually transferred to the uterus simultaneously, there is a heightened risk that the mother will carry multiple fetuses, which can make for a higher risk pregnancy for both the mother and fetuses. In addition, as with IVF generally, there is no certainty that a pregnancy will occur after the embryo is transferred. One known risk specific to PGD is that the biopsy to remove one or two cells from the embryo for genetic testing may harm or destroy the embryo.

As for the accuracy and effectiveness of genetic testing, currently there is no government review of the analytic or clinical validity of a genetic test before it is marketed. There have been a small number of cases in which PGD failed to detect the genetic abnormality it was intended to reveal. The targeted condition was later detected either during pregnancy or following the birth of the child. Because an error can be made when testing the embryo, it is often recommended that the PGD result be confirmed by subsequent prenatal tests, such as amniocentesis or chorionic villus sampling. Some recent data suggest that PGD may increase the success rate of IVF if it is used to test embryos for chromosomal aneuploidy, but opinions vary on whether and under what circumstances this is useful.

III. CURRENT OVERSIGHT AND POLICY APPROACHES

There is currently very little oversight of PGD. In general, decisions about PGD are left to IVF and PGD providers, who, together with patients, determine if PGD is appropriate in particular situations. Though most existing oversight is indirect and enforced to varying degrees, there are several oversight entities that could potentially play larger roles in the future.

Federal Agency Oversight

PGD sits at the intersection of two technologies with ambiguous and complex regulatory status: assisted reproduction (IVF) and genetic testing. The federal government does not typically directly regulate the practice of medicine, leaving such oversight to the states. Nevertheless, there are a variety of mechanisms that governmental agencies can use to regulate or influence the safety and availability of health care services and medical products. Three federal agencies within the U.S. Department of Health and Human Services oversee areas related to PGD: the Centers for Disease Control and Prevention (CDC), the Food and Drug Administration (FDA) and the Center for Medicare and Medicaid Services (CMS, formerly known as the Health Care Financing Administration). Federally-funded research is also subject to federal regulations for the protection of human subjects of research.

The CDC oversees the 1992 Fertility Clinic Success Rate and Certification Act (FCSRA).[20] This law requires that IVF clinics report pregnancy success rates annually to the federal government. The FCSRA requires clinics to report a variety of data, and noncompliance results in the clinic being listed on the CDC's website (other than this minimal chastisement, there are no penalties for failure to comply with the law). However, the FCSRA does not require IVF clinics to report the health status of babies born as a result of the procedure or the use of diagnostic tests such as PGD.

The FDA, under the Federal Food, Drug, and Cosmetics (FD&C) Act,[21] regulates drugs and devices, including those used in IVF treatments.[22] Depending on the type of product, the FDA may require submission of data from clinical studies (premarket review) and agency approval before the product may be sold. FDA does not regulate most genetic tests, although it does regulate certain components that laboratories use to make those tests.[23] Thus, for genetic tests in general, and those used in PGD, there is no uniform system to assure accuracy or validity before the tests are marketed.

FDA also regulates facilities handling human tissues intended for transplantation in order to ensure the safety of the tissue supply. Recently, FDA has decided to extend this limited regulatory oversight to facilities handling reproductive tissues under certain circumstances.[24] In addition, FDA regulates the safety and effectiveness of certain human tissue-based therapies, such as tissues that are manipulated extensively or are used in a manner different from their original function in the body.[25] These "biological products" may be subject to premarket review and approval under the Public Health Service Act[26] and FD&C Act. However, FDA has not determined that reproductive tissues are biological products when used for IVF or PGD procedures, and it has not required premarket review or approval for these tissues. Whether FDA has the legal authority under current statutes to categorize reproductive tissues in this way or require premarket review, and whether FDA would choose to do so if legal authority were not an issue, is an open question.

The Center for Medicare and Medicaid Services implements the Clinical Laboratory Improvement Amendments of 1988 (CLIA).[27] CLIA includes requirements addressing laboratory personnel qualifications, documentation and validation of tests and procedures, quality control standards, and proficiency testing

[20] Fertility Clinic Success Rate and Certification Act of 1992.

[21] Federal Food, Drug, and Cosmetic Act.

[22] For example, medicines used to stimulate ovulation are classified as "drugs" subject to the FD&C Act and therefore must be approved by FDA before they are marketed in the United States. Similarly, culture media used to grow human embryos in the laboratory prior to implantation are classified as "devices" subject to premarket approval or clearance.

[23] Department of Health and Human Services, Federal Drug Administration, "Medical Devices."

[24] Department of Health and Human Services, Federal Drug Administration, "Human Cells, Tissues, and Cellular and Tissue-Based Products."

[25] Food and Drug Administration, "Proposed Approach," 12-17.

[26] Public Health Service Act. The biologics provisions of the Act are codified at 42 U.S.C. § 262.

[27] Clinical Laboratory Improvement Amendments.

to monitor laboratory performance. CMS has not taken a position regarding whether laboratories engaged in IVF meet the statutory definition of "clinical laboratories." CMS has similarly not taken a position regarding whether laboratories that engage in the genetic analysis component of PGD are subject to regulation as clinical laboratories. If CLIA were applied and enforced with respect to genetic analysis of preimplantation embryos, laboratories engaged in this activity would be required to demonstrate proficiency under CLIA's general proficiency testing requirements for high complexity laboratories. However, CMS has not yet established specific proficiency testing requirements for molecular genetic testing. Thus, the responsibility to ensure testing proficiency for genetic tests rests squarely with the individual laboratory.

Finally, research carried out at institutions supported with federal funds is subject to federal requirements for protecting human research subjects.[28] These requirements also are mandatory for research to support an application to FDA for product approval.[29] As it now stands, any research on PGD techniques involving human subjects would probably fall outside federal requirements for protecting human research subjects. First, there is a law against providing federal funding for research in which embryos are created or destroyed. Second, embryos are not generally thought to be "human subjects" within the meaning of federal regulations. Third, FDA does not currently require premarket approval for PGD.

State Regulation

No state has enacted laws that directly address PGD. Some states have passed laws related to assisted reproductive technology (ART), of which IVF is a major component. These statutes are mainly concerned with defining parentage, ensuring that the transfer or donation of embryos is done with informed consent, or ensuring insurance coverage for fertility treatment. Some states prohibit the use of embryos for research purposes and one state, Louisiana, prohibits the intentional destruction of embryos created via IVF.[30] For the most part, states have not assumed oversight responsibilities for fertility clinics.

In terms of clinical laboratories, most states do not specifically oversee laboratories that conduct IVF or PGD as part of their administration of the CLIA program. However, New York is in the process of developing standards for laboratories that will include oversight of the genetic tests associated with PGD.

Court Action and Legal Precedent

Courts have addressed a variety of cases relating to assisted reproduction but only a few concerning PGD. In one case, the parents of a child born with cystic fibrosis (CF) following PGD sued those involved with the embryo screening for failing to

[28] Code of Federal Regulations, Title 45, Part 46.
[29] Code of Federal Regulations, Title 21, Parts 50 and 56.
[30] State of Louisiana, La. Rev. Stat.

detect the condition.[31] The parents made the claim of "loss of consortium," meaning the loss of the companionship they would otherwise have had with a healthy, non-CF-afflicted child. The court construed their claim as one for "wrongful birth" and rejected it, finding that the alleged harm was too speculative. The court similarly rejected the child's claim of "wrongful life" (made by the parents on behalf of the child), which alleged that the defendants' negligent failure to detect CF denied his parents an opportunity not to give birth to him. The court also found that the defective gene itself, not the physicians, had caused the defect. Most courts that have considered the issue have rejected wrongful life claims, such as those arising from a flawed prenatal test. Part of the reason for the reluctance to accept wrongful life as a cause of action is the concern that to do so would give credence to the argument that there can be instances in which an impaired life is worse than no life at all. As more people take advantage of the new PGD technology, additional legal questions may be brought before the courts, leading to the development of a body of case law. Standards developed through case law frequently influence legislative action or become a *de facto* policy by themselves.

Self-Regulation by Professional Organizations

Medical and scientific professional organizations present another opportunity for oversight of PGD. These groups can educate their members about advances in the field, develop guidelines addressing appropriate conduct or practices and impose standards of adherence that are a prerequisite for membership. For the most part, however, such standards are voluntary: an individual can choose not to belong to the organization, and she can therefore avoid the obligation to follow its standards. Professional organizations also typically do not have authority to sanction members for noncompliance. Unless the organization is specifically authorized by the federal government to act on the government's behalf in administering and enforcing government standards, actions of the professional organization do not have the force of law.[32]

For example, the American Society for Reproductive Medicine (ASRM) is a professional organization whose members are health professionals engaged in reproductive medicine. ASRM issues policy statements, guidelines and opinions regarding a variety of medical and ethical issues that reflect the thinking of the organization's various practice committees. In 2001, ASRM issued a practice committee opinion on PGD, which stated that PGD "appears to be a viable alternative to post-conception diagnosis and pregnancy termination."[33] It further states that while it is important for patients be aware of "potential diagnostic errors and the possibility of currently unknown long-term consequences on the fetus" from the biopsy procedure, "PGD should be regarded as an established technique with specific and expanding applications for standard clinical practice." ASRM has also

[31] *Doolan* v. *IVF America.*

[32] This issue becomes more acute when there is limited legislation, as is the current situation.

[33] American Society for Reproductive Medicine, 1-4.

issued an ethics committee opinion cautioning against the use of PGD for sex selection in the absence of a serious sex-linked disease.[34]

Two other professional organizations, the PGD International Society and the European Society for Human Reproduction and Embryology (ESHRE), are potentially in a position to play a larger oversight role for PGD. ESHRE recently issued "best practice" guidelines for PGD.[35] These guidelines are an attempt to build consensus regarding PGD practice, while recognizing that different countries may adopt different practices because of country-specific requirements or circumstances. The ESHRE guidelines explicitly acknowledge that they are not intended as rules and not enforceable, but note that, in some countries, best practice guidelines may become "standard of care" and may be codified in law.

Several professional organizations oversee clinical laboratories and could potentially extend their oversight to the laboratory component of PGD. For example, the College of American Pathologists has been empowered by the federal government to inspect laboratories seeking certification under the Clinical Laboratory Improvement Amendments, and the American College of Medical Genetics develops laboratory standards or clinical practice guidelines for genetic tests. However, neither group has developed guidelines and standards for PGD.

IV. POSSIBLE OVERSIGHT ACTIONS, REVISIONS AND CONSIDERATIONS

With this plethora of possible regulatory bodies in mind, we turn to the question of what entities might oversee PGD, and what objectives such entities might seek to accomplish. We consider oversight at the federal and state level, as well as oversight through the actions of professional organizations. In the absence of agreement, decisions about when and whether to use PGD will continue to be made largely by providers and patients.

Policy Options

There are several policy approaches that could be taken to restrict the use of PGD to those purposes deemed acceptable. Federal or state legislatures could enact a law clearly prohibiting PGD for uses it determines to be unacceptable (e.g. sex selection for non-health-related uses), an approach that has been used by a number of other countries.[36] This approach would require that lawmakers list and define prohibited

[34] American Society for Reproductive Medicine, Ethics Committee.

[35] European Society for Human Reproduction and Embryology.

[36] For example, under French law PGD is allowed only if the relevant hereditary predisposition has previously been demonstrated to exist in the parents or in one parent, and, only for the purpose of avoiding a severe genetic pathology (Loi no 94-654 du 29 juillet 1994 «relative au don...»). Similarly, in the U.K. access to PGD "is confined to individuals having a known family history of a serious genetic disorder" (United Kingdom Parliament, Human Fertilisation and Embryology Authority, and Advisory Committee on Genetic Testing). Finally, Canada recently enacted a law

uses, but it also requires some degree of moral consensus among legislators and their constituencies.

Alternatively, federal or state legislatures could pass legislation delegating to a new or existing federal agency the authority to oversee PGD, for which there is also precedent from other countries.[37] For example, Congress could pass a law giving a new or existing federal agency the authority to oversee PGD. The agency would be empowered to deal with matters of safety, accuracy and effectiveness. Such an entity could be charged with: (1) Licensing and inspecting facilities that engage in PGD; (2) Approving new PGD tests and techniques; (3) Developing regulations concerning how PGD should be conducted, focusing on quality assurance and control; and (4) Collecting data on health outcomes of children born following PGD.

One statute that could be broadened to explicitly cover PGD is CLIA, which currently provides limited reassurance to patients about the safety and accuracy of genetic tests in general and PGD in particular. CLIA could be applied and enforced with respect to genetic analysis of preimplantation embryos. Proficiency testing standards for this genetic analysis could be developed. Similarly, the government could authorize FDA to broaden its oversight to require that PGD providers demonstrate the safety and effectiveness of the genetic testing and embryo biopsy components of PGD. Those IVF-PGD clinics would be required to conduct controlled clinical trials and submit data from those trials to FDA. Finally, CDC's reporting obligations for IVF clinics could be rigorously enforced and expanded to include health outcomes of children born as a result of PGD.

At the state level, public health agencies could play a role in monitoring and improving the safety and accuracy of PGD. It is, however, difficult to create a uniform policy approach for state public health agencies because they take so many different statutory and bureaucratic forms. Nonetheless, each agency could take its basic charge to protect the public health and apply it to improving the safety and accuracy of PGD, as New York State, for example, is doing for laboratory standards (as mentioned previously).

Professional organizations could provide significant oversight of PGD in ways that do not require the involvement of federal or state authorities. Non-governmental approaches to regulate PGD could involve the development of professional guidelines through a new or existing professional organization (such as the ASRM). While such guidelines are traditionally voluntary, they can nevertheless exert

---

prohibiting PGD for sex selection in the absence of a sex-linked disease-causing mutation (Parliament of Canada, Bill C-6).

[37] For example, in 1990 the U.K. enacted the Human Fertilisation and Embryology Act, which established the Human Fertilisation and Embryology Authority (HFEA). This Authority has the responsibility to license and monitor clinics that carry out in vitro fertilisation (IVF) and donor insemination, license and monitor research centres undertaking human embryo research, regulate the storage of gametes and embryos, produce a Code of Practice which gives guidelines to clinics about the proper conduct of licensed activities, maintain a formal register of information about donors, treatments and children born as a result of those treatments, provide relevant advice and information to patients, donors and clinics, review information about human embryos and any subsequent development of such embryos, and the provision of treatment services and activities governed by the HFE Act, and advise the Secretary of State on relevant developments in treatments and research (United Kingdom Parliament, "Human Fertilisation and Embryology Act").

influence on medical practice—indeed, they are often viewed as evidence of the standard of care within a particular specialty. Since guidelines are more useful when some enforcement mechanism is contemplated, perhaps membership could be contingent upon adherence to the guidelines. In addition, the organization could encourage patients and those paying for PGD services (including employers and insurance companies) to use only the services of organization members, creating market forces in favor of compliance. The professional society could give this mechanism additional authority through a campaign educating the public and payors about the benefits of using providers who are members. Since a number of different types of professionals are involved in providing PGD services (physicians, geneticists, embryologists, technicians), collaboration among several existing professional organizations would be optimal. These complementary organizations could develop a comprehensive system to certify PGD providers in clinics and laboratories, and thereby ensure a minimum level of competency. Organizations that could collaborate to develop such a system include the American Board of Medical Genetics, the American Board of Obstetrics and Gynecology, and the American Association of Bioanalysts.

A second non-governmental approach would be to employ education, rather than regulation, to discourage PGD for purposes deemed unacceptable. Patient groups, which typically are organized around particular diseases or conditions, could develop their own recommendations for appropriate use of PGD. In addition, patient groups could educate genetic counselors and other health care professionals by including the perspective of those living with the genetic disease or condition. Further, prospective parents could have the opportunity to meet with persons living with a particular condition or disability as well as their families.

There are advantages and disadvantages to each of these approaches. Congressional intervention provides the strongest potential for national uniformity and adequate enforcement, but risks treading on the practice of medicine, a traditional province of state oversight. It can also be a blunt instrument for dealing with complex and rapidly changing technologies. Federal oversight can provide scientific expertise and greater assurance that questions concerning safety, accuracy, and effectiveness are being adequately addressed, but it also tends to limit the pace of scientific development, slows access to new technologies, and increases their cost. State oversight allows for more tailored approaches, but states may lack the resources for adequate oversight. Further, state-by-state approaches may lead to inconsistency, such that availability of services could be dependent on one's geographic location. Finally, professional oversight allows those with the most direct expertise and knowledge about the practice to craft the approach, but such an approach may be hard to enforce within the profession. It is clear that deciding what limits are appropriate, and how to implement them, will be difficult for any entity.

The Problem of Access to PGD Services

PGD is expensive. It requires IVF, which costs upwards of $10,000-$12,000. The addition of PGD can add $2,500-$4,000, bringing the total cost to approximately

$12,500-$16,000. Insurers may not cover PGD at all, or may pay only for the genetic testing, leaving prospective parents to pay for the IVF. Without coverage, PGD is available primarily to those who can pay significant out-of-pocket costs. Families who would face the greatest financial burden of caring for children born with conditions detectable via PGD may be the ones least able to afford it. Many insurers do not cover IVF for infertility treatment or offer only a limited benefit. Some fertility clinics offer ways to make IVF more affordable, and fifteen states have enacted some type of infertility insurance coverage law, but there are no federal laws in this area.

Further, there has been no systematic investigation of insurance coverage practices for PGD. No federal or state law, either enacted or proposed, requires health insurers to cover PGD. To further complicate the matter, due to a quirk in federal law, both Congress and state legislatures would have to act in order to require insurance coverage of all aspects of PGD. Anecdotal evidence suggests that when families use PGD to avoid serious genetic disorders, insurance companies are more willing to consider PGD medically necessary and cover the cost. At least one insurance company covers the genetic testing component of PGD for detection of inherited genetic disorders but not for aneuploidy. However, that company will not cover IVF if used only to perform PGD (i.e., if the IVF is not needed because of infertility).[38]

For insurers, the question of whether to cover any medical procedure or test primarily comes down to an analysis of the potential costs and benefits of coverage. A cost-benefit analysis of PGD would have to take into account the cost of the underlying IVF, the embryo biopsy and the genetic testing. It is not clear whether any health insurer in this country has undertaken a formal cost-benefit analysis of PGD for inherited genetic disorders. There could be pressure on insurers not to pay for PGD services given the moral issues involved. And from a health policy standpoint, there could be an argument made that there are many other health care needs that should be covered first.

There are several alternatives to increase access to PGD without government mandates. Private employers could include PGD in their employee benefit plans. IVF clinics and PGD providers and laboratories could offer financial assistance directly to prospective parents seeking PGD. Due to the high cost associated with assisted reproductive technologies, some IVF programs offer IVF on a "shared-risk," "warranty," "refund" or "outcome" basis. These plans operate by refunding a portion of the fee paid for one or more IVF cycles in the event that they do not result in a pregnancy or live birth of a child. Typically, shared-risk patients pay a higher fee than other IVF patients and, in return, receive a refund of 70 to 100 percent of this fee if treatment fails. While this means that someone who does have a baby may pay more under the shared-risk plan than she would have under a traditional fee-for-service plan, this option helps ensure that non-pregnant couples will have the monetary resources to pursue other options for starting a family.

[38] Aetna, http://www.aetna.com/cpb/data/CPBA0358.html.

Research and Data Collection

While it is sometimes fashionable in science and policy to opine that "more research is needed," in the case of PGD critical data are truly needed to develop effective, evidence-based policy. Research is warranted in two directions: the safety, accuracy, and effectiveness of PGD itself; and the informed public's attitudes toward this technology.

Many questions remain about the safety, accuracy and effectiveness of PGD. These include how often embryo biopsy damages or destroys embryos, how often PGD fails to detect a genetic mutation, and whether and for whom aneuploidy screening improves IVF results. In addition, more research is needed on the genetic tests used in PGD in order to improve test validity. There are incomplete and conflicting data on the long-term health effects of IVF for women and children, and no systematic studies on the health and developmental outcomes for children born following PGD. Thus, it is difficult to assess the baseline risk of IVF and any possible additional risk from the biopsy component of PGD. Longitudinal studies of women who have undergone IVF and children born following IVF and PGD would provide valuable information about the safety and risks of IVF and embryo biopsy.

Funding for such research could come from a variety of sources, including industry, private foundations and the federal government. Federal funding would, however, be limited to research not involving the creation or destruction of human embryos unless Congress lifted the current funding ban. To obtain that information, the Fertility Clinic Success Rate and Certification Act (administered by CDC, together with the Society for Assisted Reproductive Technology) could be expanded and enforced, requiring IVF clinics to report when PGD is used as part of an IVF procedure. Information required could include the purpose for which PGD was used (e.g., aneuploidy, cystic fibrosis), whether pregnancy occurred, and the outcome of such pregnancy. Currently, clinics that fail to report information on IVF procedures face no penalties. Officials could consider monetary or other penalties for failure to report. In terms of insurance coverage, further research is needed to clarify current coverage policies for PGD, the extent to which price is a barrier to patient access to PGD and the costs and benefits for third-party payors (insurance companies and employers) of covering PGD.

In addition, more empirical and theoretical research is needed on the potential societal impact of PGD. To address these questions, researchers could track changes in resources available for the disabled and in societal perceptions over time (the same could be applied to gender selection concerns). Longitudinal psychological studies of families who have used the procedure for a variety of reasons would provide data on the impact of PGD on families. And last but not least, surveys of national opinion and education about this topic will be essential in the formulation of any policy and oversight of PGD.

V. PGD AND ITS FUTURE IMPLICATIONS FOR SOCIETY

Looking to the future, some observers view PGD, or any technology that allows parents the ability to choose the characteristics of their children, as having the potential to fundamentally alter the way we view human reproduction and our offspring as well. They fear that human reproduction could come to be seen as the province of technology, and that children may be viewed as the end result of a series of meticulous, technology-driven choices.

Some argue that widespread use of PGD eventually could change the current framework of social equality in many areas. The most dramatic possibilities involve babies who are born with genes selected to increase their chances of having good looks, musical talent, athletic ability, high SAT scores or whatever a parent who can afford PGD may desire. Meanwhile, such advantages would be unavailable to the less affluent.

Such a scenario, while not possible now, is perhaps not totally implausible. Although PGD involves a diagnostic test and embryo selection, it is not genetic manipulation or "engineering" of the embryo itself. Over time the factors for which an embryo is tested could grow as science elucidates the links between individual genes and specific traits. Embryos might be selected or discarded based on genes correlated with intellectual, physical, or behavioral characteristics.

Two sources of concern regarding PGD relate to its impact on the disabled and on women. First, some worry that, with the increasing number of genetic tests and ability for embryo selection, PGD could negatively impact the way society views the disabled. Some critics argue that some of the genetic conditions that PGD can now detect, such as those causing hereditary deafness, are merely human differences that do not limit an individual's ability to live a useful and satisfying life. To select against these human differences would be to create a problematic "norm" for human flourishing. Using technology to prevent such births, these groups argue, will lead to a society in which aesthetic concerns, convenience or mere prejudice supplant the inherent dignity due to every human being, regardless of how closely he or she conforms to some ideal of normality or perfection. They worry that societal norms will evolve such that parents who are at risk of having affected children will be pressured to use PGD, even if they find the procedure objectionable.

Others have responded that for some time now parents have had the option of using amniocentesis and other types of prenatal diagnostic tests to probe for the same genetic abnormalities PGD can now detect. This information sometimes prompts parents to terminate a pregnancy to avoid having a child with a disability; yet many parents still choose to decline testing and to give birth to children with disabilities. Society continues to support families who make these choices. Nevertheless, anti-discrimination laws, public education, and social programs to aid the disabled would all help to limit negative perceptions of the disabled, and might also reduce the use of PGD by those who are concerned about the potential societal stigma of having a disabled child.

Specific concerns also have been raised about the societal impact of using PGD for sex selection, based on parental preferences and not on sex-linked genetic disease. Historically, females in many societies have been subjected to discrimination based purely on gender, and some cultures still openly prefer male children to female. Given this situation, some observers see using PGD for sex selection as having the potential to devalue women. However, others argue that in many countries, including the U.S., one sex is not currently preferred over the other. When sex selection has occurred, these proponents claim, boys and girls have been equally selected. Providers have varied policies as well: Some refuse to conduct tests that would allow for gender selection unless it is related to a genetic condition; others actively advertise sex-selection services.

Additional societal concerns have been raised about the potential for PGD to alter childhood and family dynamics, particularly when it comes to parental expectations and sibling relationships. For example, parents could end up being more critical and demanding of a child they view as having been carefully selected to possess certain attributes. Also, there could be tension among siblings when one is the product of PGD and the other is not, or when one has been selected via PGD to serve as an immunological match for another. In all cases of PGD, counseling guidelines could and should be developed that help prompt prospective parents to consider the breadth and implications of such matters.

CONCLUSION

Preimplantation genetic diagnosis raises many scientific questions and ethical quandaries. In this chapter we have reviewed the scientific, social, ethical and legal issues surrounding PGD and presented a range of policy alternatives that could be employed to address specific concerns. How then does one make policy in this complex and controversial area? A full consideration of all the policy alternatives and the benefits and burdens, and the range of persons or entities entrusted with making these decisions, must ensue. Only attentive and thorough debate over these topics will ensure that policy decisions in this arena are undertaken with a clear understanding of the potential impact of each alternative. An on-going effort to assess public attitudes about PGD and other reproductive genetic technologies will give stakeholders and policymakers a better feel for the diversity of opinion that surrounds these issues. We hope that it will also enable individuals and society as a whole to use technology wisely and to flourish.

REFERENCES

Adams, Karen E. "Ethical Considerations of Applications of Preimplantation Genetic Diagnosis in the United States." *Medicine and Law* 22, no. 3 (2003): 489-94.

Aetna. "Clinical Policy Bulletin No. 0358, Prenatal Diagnosis of Genetic Diseases." September 19, 2003. http://www.aetna.com/cpb/data/CPBA0358.html.

American Society for Reproductive Medicine. "A Practice Committee Report: Preimplantation Genetic Diagnosis." June 2001.

American Society of Reproductive Medicine, Ethics Committee. "Sex Selection and Preimplantation Genetic Diagnosis." *Fertility and Sterility* 72, no. 4 (1999): 595-8.

Botkin, Jeffrey R. "Ethical Issues and Practical Problems in Preimplantation Genetic Diagnosis." *Journal of Law, Medicine & Ethics* 26, no. 1 (1998): 17-28.

Catholic Health Association of the United States, "Genetics, Science, and the Church: A Synopsis of Catholic Church Teachings on Science and Genetics." 2003. http://www.chausa.org/transform/genscience.pdf.

Clinical Laboratory Improvement Amendments of 1988, P.L. 100-578, 102 Stat. 2903 (1988) (codified at 42 U.S.C. § 263a et seq.).

Code of Federal Regulations. Title 21, Parts 50 and 56.

________. Title 45, Part 46. Department of Health and Human Services, Regulations for the Protection of Human Subjects, 45 C.F.R. Part 46 (2004).

Delhanty, Joy A. "Preimplantation Diagnosis." *Prenatal Diagnosis*. 14, no. 13 (1994): 1217-27.

Department of Health and Human Services, Food and Drug Administration. "Medical Devices; Classification/ Reclassification; Restricted Devices; Analyte Specific Reagents. Final Rule." *Federal Register* 52, no. 225 (November 21, 1997): 62243-62260.

________. "Human Cells, Tissues, and Cellular and Tissue-Based Products; Establishment Registration and Listing. Final Rule." *Federal Register* 66, no. 13 (January 19, 2001): 5447-5469.

Doolan v. IVF America, 12 Mass.L.Rep. 482 (MA Sup.Ct.2000).

European Society for Human Reproduction and Embryology. ESHRE PGD Consortium, "Best Practice Guidelines for Clinical PGD/PGS testing." http://www.eshre.com/ecm/main.asp?lan=46&typ=122&fld=3603.

Federal Food, Drug, and Cosmetics Act, Chapter 675, 52 Stat. 1040 (1938) (as amended). Codified at 21 U.S.C. § 301 et seq.

Ferraretti, A.P., M.C. Magli, L. Kopcow, and L. Gianaroli. "Prognostic Role of Preimplantation Genetic Diagnosis for Aneuploidy in Assisted Reproductive Technology Outcome." *Human Reproduction* 19, no. 3 (2004): 694-9.

Fertility Clinic Success Rate and Certification Act of 1992. Pub. L. 102-493, 106 Stat. 3146 (1992). Codified at 42 U.S.C. § 263a-1 et seq.

Food and Drug Administration. "Proposed Approach to Regulation of Cellular and Tissue-Based Products." February 28, 1997. http://www.fda.gov/cber/gdlns/celltissue.pdf.

________. Regulations for the Protection of Human Subjects, 21 C.F.R. Part 50 and Part 56 (2004).

GeneTests. 26 March 2004. http://www.genetests.org/.

Grifo, J.A., Y.X. Tang, S. Munne, M. Alikani, J. Cohen, and Z. Rosenwaks. "Healthy Deliveries from Biopsied Human Embryos." *Human Reproduction* 9, no. 5 (1994): 912-6.

Handyside, Alan H. and Joy A. Delhanty. "Preimplantation Genetic Diagnosis: Strategies and Surprises." *Trends in Genetics* 13, no. 7 (1997): 270-75.

Handyside, A.H., J.G. Lesko, J.J. Tarin, R.M. Winston, and M.R. Hughes. "Birth of a Normal Girl After In Vitro Fertilization and Preimplantation Diagnostic Testing for Cystic Fibrosis." *New England Journal of Medicine* 327, no. 13 (1992): 905-9.

Human Cells, Tissues, and Cellular and Tissue-Based Products; Establishment Registration and Listing, 66 Fed. Reg. 5447 (Jan. 19, 2001) (final rule) (codified at 21 C.F.R. § 1271.1 et seq.).

Kass, Leon. *Beyond Therapy: Biotechnology and the Pursuit of Happiness*. Regan Books, 2003. See also President's Council on Bioethics, *Beyond Therapy: Biotechnology and the Pursuit of Happiness*. Washington, D.C. (2003), available at http://www.bioethics.gov/reports/beyondtherapy/index.html.

Kuliev, Anver and Yuri Verlinsky. "Thirteen Years' Experience of Preimplantation Diagnosis: Report of the Fifth International Symposium on Preimplantation Genetics." *Reproductive BioMedicine Online* 8, no. 2 (2004): 229-35.

Kuliev, Anver and Yuri Verlinsky. "The Role of Preimplantation Genetic Diagnosis in Women of Advanced Reproductive Age." *Current Opinion in Obstetrics and Gynecology* 15, no. 3 (2003): 233-8.

Loi no 94-654 du 29 juillet 1994 «relative au don et à l'utilisation des éléments et produits du corps humain, à l'assistance médicale à la procréation et au diagnostic prénatal».

Medical Devices; Classification/Reclassification; Restricted Devices; Analyte Specific Reagents, 62 Fed. Reg. 62243 (Nov. 21, 1997) (final rule) (codified at 21 C.F.R. §§ 809.10(e), 809.30, 864.4010(a), and 864.4020).

Middleton, A., J. Hewison and R.F. Mueller. "Attitudes of Deaf Adults toward Genetic Testing for Hereditary Deafness." *American Journal of Human Genetics* 63 (1998): 1175-1180.

Munne, Santiago. "Preimplantation Genetic Diagnosis and Human Implantation–A Review." *Placenta* 24 (2003): S70-6

Munne, Santiago and Dagan Wells. "Preimplantation Genetic Diagnosis." *Current Opinion in Obstetrics and Gynecology* 14, no. 3 (2002): 239-44.

Parliament of Canada. Bill C-6, "An Act Respecting Human Reproduction and Related Research" (Royal Assent received March 29, 2004).

Public Health Service Act. Chapter 373 (1944). Codified at 42 U.S.C. § 201 et seq.

Rechitsky, S., O. Verlinsky, A. Chistokhina, T. Sharapova, S. Ozen, C. Masciangelo, A. Kuliev, and Y. Verlinsky. "Preimplantation Genetic Diagnosis for Cancer Predisposition." *Reproductive Biomedicine Online* 5, no. 2 (2002): 148-55.

Robertson, John A. "Extending Preimplantation Genetic Diagnosis: The Ethical Debate. Ethical Issues in New Uses of Preimplantation Genetic Diagnosis." *Human Reproduction* 18, no. 3 (2003): 465-71.

________. "Extending Preimplantation Genetic Diagnosis: Medical and Non-Medical Uses." *Journal of Medical Ethics* 29 (2003): 213-216

Sermon, K., M. De Rijcke, W. Lissens, A. De Vos, P. Platteau, M. Bonduelle, P. Devroey, A. Van Steirteghem, and I. Liebaers. "Preimplantation Genetic Diagnosis for Huntington's Disease with Exclusion Testing." *European Journal of Human Genetics* 10, no. 10 (2002): 591-8.

State of Louisiana. La. Rev. Stat. 9:129 (2004).

United Kingdom Parliament. Human Fertilisation and Embryology Act, 1990, ch. 37 (Eng.).

________. Human Fertilisation and Embryology Authority, and Advisory Committee on Genetic Testing, Consultation document of PGD (2000).

Verlinsky, Yuri., A. Handyside, J.L. Simpson, R. Edwards, R., A. Kuliev, A. Muggleton-Harris, C. Readhead, I. Liebaers, E. Coonen, and M. Plachot. "Current Progress in Preimplantation Genetic Diagnosis." *Journal of Assisted Reproduction and Genetics* 10, no. 5 (1993): 353-60.

Verlinsky, Yuri and Anver Kuliev. "Current Status of Preimplantation Diagnosis for Single Gene Disorders." *Reproductive Biomedicine Online* 7, no. 2 (2003): 145-50.

________. "Preimplantation Polar Body Diagnosis." *Biochemical and Molecular Medicine* 58, no. 1 (1996): 13-17.

Verlinsky, Y., S. Rechitsky, S. Evsikov, M. White, J. Cieslak, A. Lifchez, J. Valle, J. Moise, and C.M. Strom. "Preconception and Preimplantation Diagnosis for Cystic Fibrosis." *Prenatal Diagnosis* 12, no. 2 (1992): 103-10.

Verlinsky, Y., S. Rechitsky, W. Schoolcraft, C. Strom, and A. Kuliev. "Preimplantation Diagnosis for Fanconi Anemia Combined with HLA Matching." *Journal of the American Medical Association* 285, no. 4 (2001): 3130-3.

Verlinsky, Y., S. Rechitsky, O. Verlinsky, C. Masciangelo, K. Lederer, and A. Kuliev. "Preimplantation Diagnosis for Early-Onset Alzheimer Disease Caused by V717L Mutation." *Journal of the American Medical Association* 287, no. 8 (2002): 1018-21.

KAREN LEBACQZ

CHOOSING OUR CHILDREN

The Uneasy Alliance of Law and Ethics in John Robertson's Thought

Technologies such as artificial insemination, *in vitro* fertilization, and prenatal diagnosis have simultaneously expanded the arena of reproductive choice and generated ethical controversy. The Human Genome Project and the possibility of nuclear transfer cloning both promise yet more possibilities that will impact decisions about having children. How are we to understand the new choices available to individuals and couples? Should we welcome genetic interventions, or be wary of them?

While church groups struggle to find a perspective on these new developments that call human reproduction into question,[1] lawyer John Robertson proffers a framework for thinking about new reproductive and genetic technologies.[2] Based on his understanding of the encoding of "procreative liberty" into American law and practice, Robertson presents a clear statement of the liberal position as that philosophy has developed distinctively in the West. Robertson's argument has evoked numerous criticisms;[3] my purpose is not to rehearse all of those, nor to offer a full review of Robertson's work. Rather, my focus is on the specific arena of *choosing the genetic characteristics of children.*

I will argue (1) that a "right to procreate" does not necessarily entail a "right to choose the characteristics of children," as Robertson claims that it does; (2) that Robertson's own argument forces him toward a moral arena beyond "rights" language—an arena he wants to eschew but is ultimately forced to acknowledge; and (3) that missing from Robertson's "rights" discussion is a consideration of justice. This omission is particularly striking in light of the close link between justice and rights language; it exposes in stark relief some of the values that in fact undergird Robertson's argument.[4]

[1] See, e.g., Episcopal Diocese of Washington, D.C., Committee on Medical Ethics, *Wrestling with the Future*.

[2] Robertson's position is most fully displayed in *Children of Choice: Freedom and the New Reproductive Technologies*, a book which Gilbert Meilaender calls a "summa" (Meilaender, 62). Robertson has extended his thoughts in several recent essays that largely amplify the earlier view: "Liberalism and the Limits of Procreative Liberty," "Genetic Selection of Offspring Characteristics," "Liberty, Identity, and Human Cloning," "Human Cloning and the Challenge of Regulation," "Oocyte Cytoplasm Transfers and the Ethics of Germ-Line Intervention."

[3] Two recent theological critiques are Meilaender, chapter 3; and Verhey.

[4] We solicited a response to this essay from John A. Robertson, who requested a copy of the essay but did not contribute a response to the volume. *Ed.*

A. W. Galston and C. Z. Peppard (eds.), Expanding Horizons in Bioethics, 123-139.

ROBERTSON'S ARGUMENT: "CHILDREN OF CHOICE"

Robertson's argument is both simple and subtle, both compelling and confusing. The gist of it can be stated bluntly: he argues that (1) human beings have a "right" of procreative liberty—a right to choose to reproduce or not; (2) there is a corollary right to choose the characteristics of one's offspring; and (3) there is another corollary right to use any developing technologies that increase our "choices" about having children and about their characteristics. (Thus, for example, there is a right of access to prenatal diagnosis or to *in vitro* fertilization.) Within the context of the United States, Robertson understands these three rights to be both legally enforceable and ethically warranted. He also understands them to be "negative" rights that protect the individual against interference but do not necessarily imply a "positive" right to services needed in order to reproduce.[5] (For example, one has a right of access to prenatal diagnosis, but not a right to have the government pay for such services.)

Robertson defines procreative liberty as "the freedom to reproduce (in the genetic sense) or not to reproduce."[6] The legal status of this right has been established in various laws and court cases—e.g., in *Skinner v. Oklahoma* the right to marry and procreate was declared to be among "the basic civil rights of man."[7] From this legal tradition, Robertson infers a moral basis. Procreative liberty is a primary value, he suggests, because reproduction is central to personal identity, dignity, and meaning: "Procreative liberty has a firm moral basis in the importance that reproductive decisions have for individuals."[8] Hence, "to deny procreative choice is to deny... respect and dignity at the most basic level."[9]

From this right of procreative liberty, two corollary rights follow. The first is the right to choose the characteristics of children. "Selection decisions," declares Robertson, "are essential to procreative liberty because of the importance of expected outcomes to whether a couple will start or continue a pregnancy."[10] In other words, information about the characteristics of offspring (the "makeup of the packet" as he later calls it[11]) is material to the decision of whether one will exercise one's right of procreative choice. "If a couple would not reproduce if a child had gene A but would if it had gene B, procreative liberty should protect their decision not to reproduce in the first case and to reproduce in the second."[12] Thus, freedom to

[5] Robertson, *Children of Choice,* 20.

[6] Ibid., 22. Robertson later makes a distinction between the right to reproduce *tout court* and the right to reproduce within a context of a desire to rear a child; we will return to this below.

[7] Ibid., 36. *Skinner v. Oklahoma* (1942) was a landmark Supreme Court decision that overturned compulsory sterilization, which had been approved in the famous case of *Buck v. Bell* (1927). *Buck v. Bell* dealt with mentally retarded persons; *Skinner v. Oklahoma* dealt with habitual criminals; in both cases, state laws had permitted compulsory sterilization for eugenic reasons.

[8] Robertson, "Genetic Selection of Offspring Characteristics," 425.

[9] Ibid.

[10] Robertson, *Children of Choice*, 152.

[11] Robertson., "Liberty, Identity, and Human Cloning," 1390.

[12] Robertson, "Genetic Selection of Offspring Characteristics," 427.

choose genetic characteristics of one's offspring is encompassed under the basic right of procreative liberty.

If there is a right to choose to have children or not, and a right to choose the characteristics of children, then there is another corollary right: the right to use existing technologies in the exercise of one's procreative liberty (i.e. having children, refraining from having children, and choosing the characteristics of children). Technologies such as in vitro fertilization or prenatal diagnosis[13] are "means to achieve or avoid the reproductive experiences that are central to personal conceptions of meaning and identity."[14] Hence, the right to use such technologies is also implied under procreative liberty: "if procreative liberty is taken seriously, a strong presumption in favor of using technologies that centrally implicate reproductive interests should be recognized."[15]

One has, then, in Robertson's view, three rights: the right to reproduce (or not), the right to choose the genetic characteristics of one's children, and the right to use technologies to secure these ends. These rights are presumptive and not absolute; they *can* be overridden.[16] Since reproductive liberty is a basic value, however, its primacy "sets a very high standard for limiting those rights…."[17] Specifically, Robertson asserts that where procreative liberty is centrally implicated, it can be restricted *only* when the effects of exercising it represent "substantial harm" to the "tangible interests" of others.[18] Quite deliberately and very importantly, Robertson excludes from such harms any "symbolic" concerns such as notions of what constitutes proper procreation or other acceptable behavior.[19] Religiously based repugnance at the use of *in vitro* fertilization would not be sufficient to override a woman's right to use this technology, nor would religious (e.g., Roman Catholic) concerns regarding the devaluation of the fetus in abortion constitute a 'harm' sufficient to override a woman's right to use prenatal diagnosis and selective abortion. Rather, repugnance or moral views of the status of the fetus are seen as "symbolic" concerns which Robertson understands to be contested ground in pluralistic society. Reasonable people, for example, will have differing moral perceptions about paid surrogacy: some find the practice demeaning to women while others do not.[20] Where there are such divisions, he argues, "symbolic concerns alone should not override the couple's interest in having and rearing biologic offspring…."[21] If concerns are symbolic rather than tangible individual interests or

[13] See the essay by Kathy Hudson, Susannah Baruch and Gail Javitt in this volume, entitled "Genetic Testing of Human Embryos: Ethical Challenges and Policy Choices." *Ed.*

[14] Robertson, *Children of Choice*, 220.

[15] Ibid., 40.

[16] Robertson, "Genetic Selection of Offspring Characteristics," 428.

[17] Robertson, *Children of Choice*, 30.

[18] In Robertson, "Genetic Selection of Offspring Characteristics," 428. Roberston calls it "compelling, tangible harm."

[19] Robertson, *Children of Choice*, 41.

[20] Ibid., 141.

[21] Ibid.

harms, they do not suffice to override individual freedom. Robertson's resistance to symbolic concerns continues into his later writing as well.[22]

Robertson thus articulates an approach to new genetic technologies based on liberal philosophy. In his schema, individual rights and liberties are paramount, and only serious harms to tangible interests of others will be sufficient to override them.[23] In a pluralistic society in which reasonable people disagree about moral judgments, this argument centering on shared perceptions of a central right that has been well codified in American law seems important as a framework for approaching new genetic technologies. But does the argument hold? I think not. There are at least three major problems.

I. CHOOSING CHARACTERISTICS — THE UNEXAMINED LEAP

In pluralistic society even a presumed right to procreate might be contested ground, but for purposes of argument I will accept this basic premise offered by Robertson. There certainly is encoded into American law a right of procreation whose grounds are well articulated by Robertson.[24] But does a right to procreate necessarily entail a right to choose the characteristics of our children, as Robertson asserts? Robertson's argument is that the characteristics of our children-to-be are so important to our decision as to whether or not we will procreate that choosing characteristics is implied by the right of procreation. But there is an unexamined leap of logic here—a leap that crumbles under closer examination.

To be sure, the characteristics of my children-to-be might be important to a decision about procreation. I might not want to have another child if I already have five sons and cannot be assured that my next child will be a girl. Nonetheless, historically a right to procreate has entailed uncertain outcomes: I can choose to have a child or not, but I cannot predetermine what the characteristics of my child will be. Before the days of amniocentesis, prenatal diagnosis, and the idea of "designer genes," "destiny" was the reproductive rule. People accepted whatever fate (i.e., genetics and reproductive physiology) brought them. One could choose *not* to have children, utilizing various means, but if one chose *to* have children, one then took one's chances. Historically, then, there is no *necessary* link between a right to procreate and a right to choose characteristics of children.

What has changed, of course, is that technology now makes it possible to give people information about the developing fetus. On the basis of prenatal information, people can choose whether to continue a pregnancy or to abort. For example, prenatal diagnosis can tell a woman whether her child will have Down's syndrome; she may choose to abort if she believes that she cannot raise a child with this

[22] In "Human Cloning and the Challenge of Regulation" for example, Robertson argues that a ban on cloning would serve largely "symbolic" purposes and that symbolic legislation often has substantial costs.

[23] In this short essay, I cannot do full justice to the subtleties of Robertson's argument. For a more extensive treatment, See McLean, "Of Genes and Generations."

[24] Robertson, *Children of Choice*, 36-37.

genome. Making such decisions is now common practice in the United States.[25] Thus, we do routinely support efforts to obtain and use genetic information in making reproductive decisions. Robertson is therefore correct to suggest that "principles of reproductive freedom and family autonomy appear to support a presumptive liberty right to obtain and use genetic information in making reproductive decisions."[26] But does this information and procreative liberty carry with it a new reproductive right to choose characteristics of one's offspring?

A right to obtain and use information is not the same thing as a right to choose the characteristics of our children. To say that one can choose *not* to have a child with a particular disorder, such as Down's syndrome or Huntington's disease, is not at all the same thing as saying that one can choose to *have* certain characteristics in one's children. Choosing *not* to have A is not the same as choosing to *have* B, C, or D. The crux of the matter is that choosing against certain known characteristics does not imply that one can choose *every* characteristic of one's child. The development of predictive technologies changes the circumstances in which we act, but does not of itself bring new rights. As Robertson himself acknowledges, "the idea of shaping or engineering offspring traits seems to be an unprecedented exercise of control over the lives of others."[27]

Robertson thus finds himself on the horns of a dilemma. It is a dilemma caused in part by conflating *the right to have information on which certain (limited) choices can be made* into *the right to make any and all possible choices*. These two rights are not the same. The dilemma emerges through Robertson's assumption that a right to have children entails a right to choose their characteristics; but this is a logical leap that Robertson has not succeeded in traversing. A right to have to children does not necessarily entail a right to choose characteristics. Nor is there *de facto* a right to choose characteristics simply because the technology is available to do so. The articulation of any such rights must be predicated upon an understanding of the meaning and purposes of having children, as Robertson himself later comes to acknowledge.[28] But this means that there are ethical assumptions underlying Robertson's argument—assumptions that deserve examination.

II. THE SILENT PARTNER: ETHICAL ASSUMPTIONS

Since the characteristics of children may be central to a decision to reproduce, and since procreation is central to personal meaning and identity, then it follows for Robertson that whatever our anxieties may be about the control we begin to exercise over others' lives, we do have a right to choose the characteristics of children.

Or do we? In fact, Robertson is not comfortable with every implication of his own logic here. With the advent of the Human Genome Project, for example, we

[25] For an interesting account of the social forces brought to bear on a woman who chose to carry a Down's syndrome fetus to term, and her own interpretation of pregnancy and motherhood, see *Expecting Adam* by Martha Beck.

[26] Robertson, "Genetic Selection of Offspring Characteristics," 422.

[27] Robertson, "Genetic Selection of Offspring Characteristics," 423.

[28] Robertson, "Liberty, Identity and Human Cloning."

may someday have the ability to choose characteristics such as blue eyes or curly hair. If we have a right to choose the characteristics of our children, then by extension, in Robertson's system we have a right to make these choices. Yet Robertson hesitates here. He acknowledges that some parents may want to choose characteristics based not on health considerations but on personal preferences (such as sex or eye color).[29] Robertson tries to avoid the implication of the logic of personal choice by suggesting that such characteristics are not "central" or "material" to reproductive decisions. But as Meilaender points out, "what is peripheral to one person's self-defining experience may be quite central to another's...."[30] While Robertson's *language* seems to argue for freedom of choice, his argument actually seeks to constrain choices within the rubric of accepted social practices in Western culture, in which certain assumptions define acceptable parenthood. Hence, he hesitates to condone selection of traits such as sex, curly hair or eye color.

Under the guise of supporting free choice, then, Robertson imports a substantive notion of parenthood. He assumes that what parents will choose is "normal, healthy offspring," to whom one is related genetically.[31] This—and only this—is what Robertson really intends for the right of reproductive liberty. Anything that deviates from the standard of a healthy child to whom one is related genetically is not intended to be encompassed by his reproductive rights. For example, Robertson argues that nontherapeutic genetic enhancement "will not be protected by procreative liberty because these actions conflict with the values that undergird respect for human reproduction."[32] Only those actions that are consistent with a certain range of values will be protected by Robertson's reproductive freedom.

If reproductive liberty is to encompass only actions consistent with certain values, then the notion of procreative freedom clearly imports what Robertson elsewhere has labelled "symbolic" concerns. Robertson is forced to move beyond the language of rights *simpliciter* into a larger arena in which rights are given shape and content by their place in a societal system of expectations and values. This is precisely the arena that Robertson has elsewhere tried to eschew by rejecting "symbolic" values in discussion of proper use of reproductive technologies. *Contra* Robertson, I will show that symbolic or value concerns always define what constitutes an acceptable reproductive liberty.

[29] It should be noted that some chronic and terminal conditions are sex-linked. In this section I am dealing only with the question of sex selection based on preference.

[30] Meilaender, 68.

[31] Robertson later adds that the individual or couple must also desire to *rear* the child themselves. See "Liberty, Identity, and Human Cloning," 1393, where cloning is not protected by the right of procreative liberty "if no rearing is intended."

[32] Robertson, *Children of Choice*, 172. In later essays, Robertson appears to move closer to acceptance of *all* enhancement technologies. He acknowledges that cloning may be the first important technology to focus not on screening out unwanted traits but on choosing positively to have certain traits. While in his earlier work, he was resistant to cloning, as will be seen below, he later comes to accept it. See Robertson, "Human Cloning and Challenge of Regulation," 120: "the resort to cloning is similar enough in purpose and effects to other reproduction and genetic-selection practices that it should be treated similarly. Therefore a couple should be free to choose cloning...."

In Robertson's theory *as stated,* only substantial harm to the tangible interests of others may override procreative rights. Values such as the moral offense someone may feel when confronted by another's actions are not allowed to count as substantial harm to one's tangible interests. Symbolic concerns for proper modes of reproduction (e.g., the argument made by Meilaender and others that *in vitro* fertilization or other technologies that separate making love from making babies are somehow wrong in themselves[33]) are not sufficient to override another's procreative liberty. Yet confronted with the possibility that parents might choose curly hair or blue eyes, or choose to enhance the characteristics of a child (e.g., to make the child taller, a trait that has distinct advantages for men in Western culture), Robertson finds such choices not "material" or sufficiently close to the "core values" that give meaning to reproductive rights. What are such "core values" if not symbolic concerns? In fact, symbolic meanings such as notions of what makes reproduction "human" become central to determining the scope of reproductive liberty in Robertson's argument.

This is clearly evident in Robertson's shifting arguments about cloning. In *Children of Choice,* Robertson suggested that cloning would not be covered by the right of procreative liberty, since "reproducing with a cloned embryo may deviate too far from prevailing conceptions of what is valuable about reproduction to count as a protected reproductive experience."[34] In his more recent work on cloning, however, Robertson suggests that, so long as rearing the child is intended, "a person's procreative liberty should... include the right to clone oneself."[35] These arguments indicate that normative judgments about what constitutes the core of parenting are absolutely essential to the development of Robertson's thought. In his early work he took for granted that reproductive rights are exercised by couples who desire to bear *and rear* children; in his later work, the expectation for rearing emerges as an assumption that must be spelled out, in light of the fact that cloning and other technologies would allow one to contribute to the genes of a child without any intention of rearing that child oneself.

In short, although Robertson wants to prevent symbolic issues such as value disagreements from having sufficient force to override procreative liberty, he is finally forced to recognize that it is precisely such values that must function to limit certain procreative choices and justify others. "Freedom to choose" does not imply a limit on that choice, nor does it imply any normative content to the choice. If I have a right to choose *simpliciter*, then I have a right to choose whatever I may desire. All limitations on choice imply some substantive standards that might justify the limitations. In other words, if limits are to be set on what can be chosen, those limits do not derive from the freedom to choose itself, but from substantive standards. Such standards are always symbolic, socially embedded understandings. Robertson's language and the logic of the liberal arguments that undergird his

[33] Meilaender.

[34] Robertson, *Children of Choice*, 169. Cf. Robertson, "Genetic Selection of Offspring Characteristics," 438: cloning "might fall outside the bounds of reproduction as commonly understood in today's society."

[35] Robertson, "Liberty, Identity, and Human Cloning," 1402.

position force Robertson toward conclusions that he does not want to accept—for example, that parents should be free to "enhance" their children, or to choose characteristics such as sex or eye color. Since what is central or material to one person's decision will differ from what is material to another's, there is no standard by which to choose what range of genetic intervention is acceptable. On the force of the argument alone, it would appear that parents should have unlimited choice over the characteristics of their children. Yet Robertson does not accept this claim.

Robertson acknowledges that "for some parents, choice of [characteristics such as] offspring gender will be central to reproductive choice, and therefore presumptively protected against restriction...."[36] If gender is sometimes central to reproductive choice, then on what grounds would one restrict a couple's freedom to choose the gender of their children? Meilaender observes that he "cannot find a thread in Robertson's argument" strong enough to establish legitimate distinctions for acceptable and unacceptable choices.[37] I concur. This is because Robertson depends for his argument almost entirely on his assessment of the current *legal* and *social* climate, and presents little if any *ethical* argument. As the social climate changes (e.g. as the thought of human cloning becomes more acceptable), so must Robertson's permissible range of parental choices change. Indeed, this is precisely what has happened with regard to Robertson's arguments about cloning.

Robertson tries to avoid the infinite expansion of personal choice by arguing that we must *combine* subjective preferences and standards with "objective" standards based on general views in the culture.[38] Choice is not the only value, and choice can be limited by shared general views. The normative judgment would then be drawn from "prevailing conceptions" or what is "commonly understood."[39] At several points, therefore, Robertson recognizes that a normative judgment lies beneath his stress on procreative liberty: "In the final analysis," he suggests, "one's view about the importance of procreative liberty comes down to a normative judgment about the importance of reproductive experience in people's lives."[40] Later he acknowledges that the "meaning of basic values such as identity, liberty, parenting, kinship and family" is being constructed "in the very course of determining" which uses of technologies are acceptable.[41]

It is not freedom alone that drives Robertson's logic, but the *meaning* of key concepts such as parenthood. This meaning emerges bit by bit in Robertson's own developing thought, as that thought reflects advances in court cases and social acceptance of new technologies. To deal with meaning, however, is always to deal in the symbolic realm; such meaning is always socially and symbolically grounded. This is a realm where ethical argument, not just sociological data regarding accepted practices, is needed. Indeed, to allow any limitation on choice is to import a

[36] Robertson, "Genetic Selection of Offspring Characteristics," 434.
[37] Meilaender, 68.
[38] Robertson, "Genetic Selection of Offspring Characteristics," 430.
[39] References to prevailing conceptions or to "our social and legal landscape" abound in *Children of Choice.* See, for example, 16, 36, 37, 41, 154, 169.
[40] Robertson. "Liberalism and the Limits of Procreative Liberty," 236.
[41] Robertson, "Oocyte Cytoplasm Transfers," 212.

substantive standard of what counts as acceptable goals and forms of reproduction—the very "symbolic" issues that Robertson elsewhere rejects as sufficient to limit reproductive freedom! Even to allow *no* limitations on choice is to make a *substantive* decision regarding the centrality of choice in human life. Procreative liberty is based on an understanding of the centrality of certain rights, but "rights" language cannot stand alone, because rights are always embedded in a social context from which they get their meaning, strength, and limitations. Robertson cannot avoid what he calls symbolic concerns; such normative and value issues are the very foundation for rights.

III. RIGHTS AND JUSTICE — THE MISSING PIECE OF THE PUZZLE

For Robertson, the grounds for choosing "objective" standards appear to be what is in common practice or has been codified into law. It is Robertson's grounding of his argument in current law and social practice that brings me to my third objection. Precisely because we are dealing with symbolic and normative issues, common practice or legal sanction may not be a good standard for determining what should be allowed. If prevailing conceptions determine what is included in procreative liberty and what is excluded from it, how do we establish the "norm" or standard of human function and human reproduction? What happens to minority groups or those with views that do not fit the prevailing conceptions?

For instance, genetic counselors report that some deaf parents want to have only deaf children—children who can be raised in Deaf culture and society.[42] If parents are free to choose the characteristics of their offspring, then deaf parents should be free to choose deaf children, even though most people consider deafness a disability or defect. Are such alternative concepts that do not fit prevailing norms allowed in Robertson's schema? Since Robertson in general tries to craft an argument that can accommodate pluralistic views, it is to be expected that he would make room for alternative conceptions of what constitutes good parenting, and that he would allow deaf parents to choose to bear children who are deaf and reject fetuses that will develop into hearing children (just as hearing parents are allowed to choose fetuses that will develop into hearing children and reject fetuses that will develop into deaf children).[43] The logical implication of a right to choose the characteristics of one's children is the right to choose children with characteristics that others would consider disabilities.

[42] See, for example, "Deaf Culture, Cochlear Implants, and Elective Disability" by Bonnie Poitras Tucker, 7. Following general usage, I use the term Deaf to indicate the politically active community that argues that deafness is not a disability. Not all deaf people are members of the Deaf community, but many are. For a discussion of the arguments made by Deaf spokespersons, see "Deafness as Culture" by Edward Dolnick. See also *The Mask of Benevolence* by Harlan. Tucker takes issue with the view that deafness is not a disability.

[43] It should be noted that deafness has a complicated etiology, and is not generally the result of a monogenic disorder that could be easily detected by prenatal diagnosis. However, some forms of deafness such as Usher's syndrome are of genetic origin.

Some, of course, would want to argue that a child deliberately created deaf does in fact suffer substantial harm to her or his tangible interests. For example, Arana Ward points out that the average deaf person reads at a fourth grade level and earns 30% less income than the general population; furthermore, deaf people are often concentrated in manual jobs with little hope of advancement.[44] Surely substantial loss of income and diminished educational and vocational opportunities might count as "substantial harm" to "tangible interests." Does the decision to have children with a particular characteristic—in this case, deafness—therefore present substantial harm to another's tangible interests?

In *Reasons and Persons,* Derek Parfit offered a case for ethical deliberation in which a woman is told that if she conceives a child while taking a certain medication, the child will be born with a withered arm; if she waits until she is off the medication, she can expect to have a non-injured child. Parfit argued that if she does not wait, she has injured the child and should be morally condemned.[45] However, Robertson disagrees. Drawing on the legal tradition that there is no grounds for "wrongful life" where life itself is at stake, Robertson argues as follows: since *this* child would not have been conceived if the woman waited several months (she would then produce a different egg, which would be fertilized by different sperm, resulting in a different child), *this* child has no grounds for complaint about its withered arm.[46] By extension, we can presume that Robertson would dismiss arguments to the effect that a deaf child, deliberately so conceived, has a claim for complaint about her or his deafness, whatever the social concomitants and tangible financial losses might be.[47]

Indeed, Robertson's line of reasoning brings him to conclude that one may knowingly and deliberately conceive a child who will suffer harm from the circumstances of conception, if that is the only way to conceive *this* child. One need not, then, in Robertson's argument, choose to have only normal, healthy children, although choosing normal and healthy children is generally his implicit and explicit norm. Ironically, of course, since Robertson has argued *against* genetic enhancement, it now appears that, following his reasoning, one can choose damaged children but not superior children! Such a conclusion seems to fly in the face of normative assumptions that parents are to try for the "best" for their children. Based on the logic of argument alone, the conclusions are bizarre at best.

[44] Ward, quoted in Tucker, 13.

[45] Parfit, 352-379.

[46] Robertson, *Children of Choice*, 76.

[47] Whether Robertson would allow financial burdens to *others* to count as substantial harms to tangible interests sufficient to override the right of procreative liberty, including the right to choose a child who is deaf, is not clear. Tucker argues that the costs to society of incorporating deaf people into social and cultural institutions is sufficient to override the freedom of parents to choose against cochlear implants.

The "Disability" Critique

But there is another and more important concern at stake. Precisely at issue is the question of whether deaf children should be considered "damaged" or "disabled" at all. Robertson's argument allows parents to choose a disabled child, and therefore would allow deaf parents to choose to have a deaf child, but it offers no challenge to the conceptualization of deaf children as disabled. By contrast, the Deaf community argues that deafness is *not* in itself a disability but is constructed as such by dominant culture.[48] The distinction between deafness as disability—even a disability that can be chosen by parents—and deafness as simple difference, but not disability, raises some fundamental issues of justice.

Who determines what concepts of health, disease, normality, and parenting will set the stage for the range of rights to be encompassed under Robertson's procreative liberty? If actions are to be protected against restriction when they are close enough to some "core" notion of reproduction,[49] then the determination of that "core" notion is all-important. In her heartbreaking study of the effects of poverty in Brazil, for example, Nancy Scheper-Hughes raises distinct challenges to Western notions of what constitutes good mothering.[50] Similarly, I would argue, the notion that children must be genetically related to us in order to be "ours," and hence that any and all efforts to secure such genetic links are justified, is a peculiarly Western notion. What we take for granted is not necessarily what would be accepted by other cultures and traditions.

Thus, although Robertson intends to extend and support individual liberty at every turn, in fact I would argue that his approach as a whole supports and even fosters the imposition of dominant views. By delineating a core of values surrounding reproduction and allowing *prevailing conceptions* to determine the range of acceptable choices, Robertson opens the door to the form of injustice that Iris Marion Young identifies as "cultural imperialism."[51] Precisely because Robertson works from within the framework of prevailing social and legal practices, he has no standpoint from which injustices built into those practices can be judged. Further, while Robertson does not tackle the question of justice directly, his understanding of justice would appear to be encompassed by equal treatment within the system. So long as everybody has the same right of procreative liberty, bounded by the same considerations of harm to the tangible interests of others and of adhering to core values of reproduction, Robertson would assume that justice has not been violated. Feminists would disagree.

[48] Lebacqz, 3-14. For a developed argument about the social construction of deafness as disability, see Lane.

[49] Robertson's argument is akin to a kind of casuistic reasoning, in which we find a 'case' or 'type' of case on which everyone agrees and then see what happens when we move further and further away from the characteristics of that case. For a developed view of casuistry, see Jonsen and Toulmin.

[50] Scheper-Hughes.

[51] Young, chapter 2: "Five Faces of Oppression."

Feminism, Justice and Rights

Robertson dismisses feminist concerns about the impact of practices of *in vitro* fertilization, surrogate mothering, and other new reproductive technologies. He recognizes that rights-based approaches have come under attack in recent years.[52] He acknowledges the feminist critique that women's control over reproduction may actually be lessened by the development of all these new technologies.[53] But he simply asserts in rebuttal that new technologies increase options that expand the freedom of women, and he takes the rights-based approach to be "the best guarantee" of women's control and liberty. He claims that "when everything is considered, a strong commitment to procreative liberty will protect more than it will harm the interests of women."[54] In short, Robertson dismisses the concerns of feminists who have rejected rights as a basis for ethics, or who see justice as involving questions of group oppression as well as distribution of individual rights. Robertson operates from within a liberal framework that assumes the crucial issue is equal individual rights. Few feminists are content with such a limited liberal framework.

Nancy Hirschmann, for example, argues that the "rights" approach reflects the detached Western male stance and does not do justice to the fact that we are born into networks of people with interlocking interests. Some of our obligations are not "chosen" but are imposed by fundamental relationships such as parenthood.[55] Hirschmann proposes that rights are too narrow a frame to express the ethical dimensions of relationships such as parenthood.[56] Similar critiques of "rights" have been made by Susan Moller Okin and Iris Marion Young.[57]

Young takes the question of justice a step further. Rights language is individual language. But we are not treated simply as individuals, she suggests. We are also treated as members of *groups.* Ethnic, racial, and sexual classifications, for example, often dominate the way others respond to us and the range of abilities and opportunities available to us. Precisely because we are discriminated against *as women* or *as black* or *as native*, for example, justice requires that the ways in which groups shape our identities must be taken into account. By extension, a deaf person is not simply treated as an individual, but will be treated *as a disabled person* by dominant culture, even if that person is a member of Deaf culture and does not consider herself disabled.

There is a danger, then, in allowing procreative choice to encompass choosing the characteristics of our children: those in dominant culture will choose based on notions of disability that derive from dominant culture and do not reflect the views of marginalized or oppressed groups. The rights framework may protect individuals

[52] Robertson, *Children of Choice*, 223.
[53] Ibid., 228.
[54] Ibid., 229.
[55] For a theologically-based argument that a focus on "choice" is inadequate to ethics, see *A Feminist Ethic of Risk* by Sharon D. Welch.
[56] Hirschmann.
[57] Okin; Young.

against discrimination *within* a system, but does nothing to ensure that the underlying standards *of* the system are themselves fair and subject to ethical scrutiny. In a critique of emerging ethical discourse around genetics, Susan Wolf argues that avoiding "genetic discrimination" (e.g. ensuring that people are not deprived of life or health insurance because of their genetic status) is not enough. The real issue, she charges, is not genetic discrimination but "geneticism."[58] Like the terms racism or sexism, "geneticism" implies not simply discrimination against individuals, but an interlocking system of oppression, marginalization, and exclusion. The issue is not whether we are treated equally in terms of genetic standards; the issue is having genetic standards at all. In Wolf's view, it is precisely the notion that there *is* a genetic norm that is the problem.

The justice issue is whether and how genetic norms are constructed in society, not whether individuals are treated equally in the face of such norms (though this is of course an important consideration). *Having* the norm is the fundamental problem. Here lies a significant weakness in Robertson's claims: Precisely because he works from within the framework of prevailing social and legal practices, Robertson has no standpoint from which having such norms can be challenged.

Further Feminist Concerns

Indeed, Robertson's methodology is sexist. Not only does he give feminist arguments short shrift, as though they count for nothing, but he consistently minimizes the position of women with regard to new genetic technologies. At best Robertson's arguments show a neglect of women, and at worst a bias against women. In his recent discussion of cloning, for example, he addresses questions of objectification and instrumentalization *of the clone* but never addresses questions of objectification and instrumentalization *of the women* whose eggs and bodies are necessary for the development of human cloning.[59] He notes that a child could have as many as four "mothers," (a nuclear DNA mother, an egg mother, a uterus/gestational mother, and a mother of rearing), but fails to address what this means for the women involved or what understanding of "motherhood" would make such a conceptualization possible.[60] To "simplify" the social and legal complexities of cloning, he proposes that the egg donor "should not be regarded as socially significant kin at all."[61] It is ironic that in a discussion of a technique that has the potential to make *men* irrelevant for purposes of reproduction, the role and place of *women* receives no attention from one whose argument is based on procreative liberty—itself a notion that owes much of its enshrinement in law to the vigorous efforts of women during the long debates on contraception and abortion in the United States. At his worst, therefore, Robertson shows a bias against women.

[58] Wolf, 345-53.

[59] Robertson, "Liberty, Identity, and Human Cloning," 1419. Elsewhere, however, he does specify that women egg donors should be "fairly treated." ("Oocyte Cytoplasm Transfers and the Ethics of Germ-Line Intervention," 215.)

[60] Robertson, "Liberty, Identity, and Human Cloning," 1427.

[61] Ibid., 1428.

Even at his best, he demonstrates a neglect and ignorance that cannot result in justice for women. For example, Robertson assumes the validity of middle-class values and perspectives that currently dominate the legal arena. His "rights" are "negative" rights—rights against interference by others—not "positive" rights that would ensure social supports to effect the exercise of liberties such as procreative liberty. But many women have little procreative liberty.[62] Absent birth control, a woman cannot simply "choose" to delay reproduction until the timing is optimal, for example. The biological, social and economic impact of simply being female, and its effect on the exercise of rights, is neglected by Robertson.[63] Without consideration of such justice issues, however, women's "rights" are an empty shell.

Similarly, although Robertson's procreative liberty is clearly based on a model of individual rights, he often switches from speaking of the individual or woman to speaking of the couple. (He also assumes a heterosexual couple, neglecting the procreative rights of lesbian women.) In addition, Robertson rarely addresses the problem of a division of opinion between the partners in a couple.[64] Some of the most poignant and difficult ethical dilemmas in genetic counseling have arisen in circumstances where one member of a couple wants to "choose" a child with a genetic anomaly or particular characteristic and the other wants to "choose" against that child. Here, the framework of individual rights cannot simply be foisted onto a partnership where there may be differences of opinion. There is also evidence that many women undergoing *in vitro* fertilization do so at considerable cost and pain to themselves because their male partners insist on having a child to whom they are genetically related, rather than adopting. To propose that the use of new technologies is simply a matter of procreative liberty, and to ignore the patterns of oppression of women in which male liberties override female choice, is not to ensure justice but to block it.

CONCLUSION

John Robertson has thrown down the gauntlet, and any future attempt to create an ethical argument in the arena of genetics and new reproductive technologies will have to take seriously his argument for *Children of Choice.* Careful analysis suggests, however, that (1) the right to choose the characteristics of our children is not necessarily implied by a basic right to procreate, (2) the right to choose does not

[62] Women's lack of "choice"—and by implication the inapplicability of rights language based on choice—is strongly argued by Welch.

[63] For a discussion of the economic consequences of being female, see *Equal Value: An Ethical Approach to Economics and Sex* by Carol S. Robb.

[64] Robertson reviews the law regarding divided opinions between couples in a short article entitled "Meaning What You Sign." In this article he supports the idea that prior directives should generally be binding: whatever a couple intended when donating gametes for IVF should serve as the principle for later disposition of embryos. His interpretation offers no sense that justice might require that those who have undergone pain and suffering in order to create embryos should have any particular claim on the embryos thus created. This contradicts Lockean notions that property or ownership rights are secured by 'mixing one's labor' with an object. For a fascinating discussion of the implications of the Lockean view, see Okin, chapter 4.

stand alone but gets its shape and articulation from a range of substantive values that need ethical articulation and argument that Robertson does not provide, and (3) it is not sufficient to depend for that shape and articulation on prevailing practices, because those practices reflect the norms of dominant culture. No argument for rights that neglects broader issues of justice, including oppression and marginalization of those who are different, will ultimately be adequate to deal with the emerging "politics of difference" in the genetic arena nor with questions of women's oppression.

Yet Robertson does us a great service by attempting to lay out the moral philosophy that currently undergirds our prevailing (and sometimes conflicting) notions about reproduction. If nothing else, perhaps he will force us to confront the flaws in our practices and logic. On the other hand, Robertson's failure to offer any challenges to those prevailing conceptions ultimately betrays the middle class and sexist biases of his work. The law may be "a repository for concepts, distinctions, principles, and precedents developed in the attempt to bring principles of justice and equity to bear on the complexity of real life,"[65] but such a repository must be critically examined in order to determine whether the concepts, distinctions, principles, and precedents embodied therein are truly serving justice. Robertson's dependence on law and failure to examine the ethical underpinnings of current legal practice leads him to an uncritical adoption of current reproductive practices. These practices do not serve well the cause of providing a just and ethical base for decisions about new genetic technologies. Robertson's thinking has developed rather haphazardly in response to the development of new technologies and legal efforts to respond to them; in order to avoid the dangers of such haphazard development, he is in need of sustained ethical analysis of the very symbolic issues that he wishes to eschew.

[65] Alpern, ed., 319.

REFERENCES

Alpern, Kenneth D., ed. *The Ethics of Reproductive Technology*. New York: Oxford University Press, 1992.

Beck, Martha. *Expecting Adam.* Berkeley Publishing Group, 2000.

Buck v. Bell. 274 U.S. 200 (1927).

Dolnick, Edward. "Deafness as Culture." *Atlantic Monthly* 272, no.3 (Sept 1993): 37-53.

Episcopal Diocese of Washington D.C., Committee on Medical Ethics. *Wrestling with the Future: Our Genes and Our Choices.* Harrisburg: Morehouse Publishing, 1998.

Hirschmann, Nancy. *Rethinking Obligation: A Feminist Method for Political Theory.* Ithaca: Cornell University Press, 1992.

Jonsen, Albert and Stephen Toulmin. *The Abuse of Casuistry: A History of Moral Reasoning.* Berkeley: University of California Press, 1988.

Lane, Harlan. *The Mask of Benevolence: Disabling the Deaf Community.* New York: Vintage Books, 1992.

Lebacqz, Karen. "Diversity, Disability, and Designer Genes: On What it Means to be Human." *Studia Theologica (Scandinavian Journal of Theology)* 50, no.1 (1996): 3-14.

McLean, Margaret. "Of Genes and Generations: Rethinking Procreative Liberty and the Right to Children of Choice." Ph.D. diss., Graduate Theological Union, 1997.

Meilaender, Gilbert. *Body, Soul, and Bioethics.* Notre Dame: University of Notre Dame Press, 1995.

Okin, Susan Moller. *Justice, Gender, and the Family.* New York: Basic Books, 1989.

Parfit, Derek. *Reasons and Persons.* Oxford: Clarendon Press, 1984.

Robb, Carol S. *Equal Value: An Ethical Approach to Economics and Sex.* Boston: Beacon Press, 1995.

Robertson, John A. *Children of Choice: Freedom and the New Reproductive Technologies.* Princeton: Princeton University Press, 1994.

________. "Genetic Selection of Offspring Characteristics." *B.U. Law Review* 76, no.3 (June 1996): 421-482.

________. "Human Cloning and the Challenge of Regulation." *New England Journal of Medicine* 339 (1998): 119-122.

________. "Liberalism and the Limits of Procreative Liberty: A Response to My Critics." *Washington and Lee Law Review* 52 (1995).

________. "Liberty, Identity, and Human Cloning." *Texas Law Review* 76, no.6 (May 1998): 1371-1456.

________. "Meaning What You Sign." *Hastings Center Report* (July-August 1998): 22-23.

________. "Oocyte Cytoplasm Transfers and the Ethics of Germ-Line Intervention." *Journal of Law, Medicine, and Ethics* 26 (1998): 211-20.

Scheper-Hughes, Nancy. *Death Without Weeping: The Violence of Everyday Life in Brazil.* Berkeley: University of California Press, 1992.

Skinner v. Oklahoma. 316 U.S. 535 (1942).

Tucker, Bonnie Poitras. "Deaf Culture, Cochlear Implants, and Elective Disability," *Hastings Center Report* 28, no.4 (1998): 6-14.

Verhey, Allen. "Commodification, Commercialization, and Embodiment." *Women's Health Issues* 7, no.3 (May/June 1997).

Ward, M. Arana. "As Technology Advances, a Bitter Debate Divides the Deaf." *Washington Post* 11 (May 1997).

Welch, Sharon D. *A Feminist Ethic of Risk.* Minneapolis: Fortress Press, 1990.

Wolf, Susan M. "Beyond 'Genetic Discrimination': Toward the Broader Harm of Geneticism." *Journal of Law, Medicine, and Ethics* 23 (1995): 345-53.

Young, Iris Marion. *Justice and the Politics of Difference.* Princeton: Princeton University Press, 1990.

HARRY M. ROSENBERG

THE HEART DISEASE EPIDEMIC THAT WASN'T

Lessons Learned from Death Certificate Statistics

INTRODUCTION

In a provocative article by Gary Taubes that examines the relationship between diet and heart disease, federal statisticians are quoted as challenging the widely held belief that an epidemic of heart disease is occurring in the United States, as follows:

> To proponents of the anti-fat message, this heart disease epidemic has always been an indisputable reality. Yet, to the statisticians at the mortality branch of the National Center for Health Statistics (NCHS), the source of all the relevant statistics, the epidemic was illusory. In their view, heart disease deaths (age-adjusted death rates) have been declining steadily since the 1940's. According to Harry Rosenberg, director of the NCHS mortality branch since 1977, the key factor in the apparent epidemic, paradoxically, was a healthier American population...[1]

In my interview with Taubes, I noted that indeed heart disease is a major problem in the United States. As the leading cause of death, it accounted for almost 725,000 out of 2.3 million deaths in 1998, that is, almost one out of three deaths. However, I stressed, despite its prevalence, heart disease in the U.S. is not *epidemic*. Strictly speaking, *epidemic* reflects a sudden upward trend in morbidity (illness) or mortality (death); a medical condition—usually infectious—that presents an imminent public health threat; and one that requires vigorous, extraordinary, and immediate measures for containment and resolution. At this point one set of ethical questions emerges: Should we call a health condition "epidemic" simply because it is a highly prevalent adverse health condition? What should be the relationship between the terminology used in describing a public health problem and the terminology used to persuade the public of needed health initiatives? Do many public health initiatives rely on the import of the term "epidemic" in order to elicit a response from the population?

Further, while proponents of the anti-fat message and I had the same data available, we chose to interpret the data in dramatically different ways. How can objectively collected and objectively disseminated public information lead to such different conclusions about the health of the U.S. population? What do we as a society need to know about health statistics to more fully evaluate the aims of

[1] Taubes, 2536-2545.

A. W. Galston and C. Z. Peppard (eds.), Expanding Horizons in Bioethics, 141-159.

disease prevention efforts? This essay addresses these questions and ultimately asserts that even widely available health statistics can be subject to misuse and deliberate misinterpretation in the interests of advocacy. Statisticians have a responsibility to identify the misuse of statistics, and to educate policymakers and the public about the use of appropriate statistical methodologies.

THE POLITICS OF HEALTH INDICATORS

Advocates like to have empirical support for their positions; and statistics, used selectively, can sometimes provide such support. When statistics are at variance with the policy objectives of advocacy groups, the very integrity of the statistics or of the statisticians that produce, analyze, and interpret the statistics may be called into question.

A compelling example of a conflict between health statistics and the policy objectives of advocacy groups occurred in 1996. At its annual meeting the President of the American Heart Association (AHA) in his presidential address severely criticized the National Heart, Lung and Blood Institute (NHLBI) for using age-adjusted death rates to portray progress in reducing the risk of heart disease in the United States. He accused the agency of exaggerating progress by selective use of statistics: "Americans have been seriously misled into thinking that heart disease is on the decline… Deaths from heart disease haven't dropped nearly as much as health officials have claimed."[2] Needless to say, the AHA President's remarks reverberated in Washington, where NHLBI—put on the defensive—established an interagency working committee with the assignment to rapidly develop educational materials that would clarify the distinction among alternative statistical indicators and their appropriate uses, and would, thereby, justify its use of age-adjusted death rates to portray progress in addressing heart disease in the U.S. Two educational pamphlets were developed by the interagency working group and were widely distributed.[3]

This example, just one of many that could be cited, illustrates the importance of choosing the appropriate statistical indicator to portray the scope of a health problem and to monitor progress in addressing that problem. Because the goal of the AHA was to portray the problem as one deserving public attention, they used simple and less sophisticated indicators that emphasized the great societal burden of heart disease (namely, the number of heart disease deaths and the crude death rate from heart disease, which will be explained below). In contrast, the federal agency wanted to depict that the "risk" of heart disease has been substantially reduced, in order to justify public investments in disease prevention programs and clinical research. Thus, the AHA emphasized the large numbers of population who have the disease and are killed by it, while the NHLBI stressed the declining likelihood of dying from heart disease to present their case to the public and policy makers. At a minimum,

[2] Bishop, B6, 2 graphs.

[3] See National Heart, Lung and Blood Institute, *Comparing Death Rates Over Time,* 1-4; and National Heart, Lung and Blood Institute, *Taking the Mystery Out of the Age-Adjusted Death Rate,* 1-4.

these contrasting statistical approaches can confuse the public and policy makers; at worst, conflicting messages about important public issues invite skepticism about the entire statistical enterprise, its methods and its practitioners, inviting the characterization that is captured so well in the phrase "lying with statistics."

Many examples can be adduced of the importance of statistics in policy making, as well as of the misuse of statistics for these purposes.[4] Few recent incidents in public health statistics, however, have been as visible and far-reaching as the bitter accusations leveled by the AHA against NHLBI in the 1990s. To understand the source of this dispute, we can look at trends in heart disease mortality in America reflected in the different types of statistical indicators widely used in public health statistics and epidemiology.

HEART DISEASE IN AMERICA: 1950-1998

Over thirty years ago, when the annual number of deaths from heart disease was increasing at about 10,000 per year, one might have dramatized the increase with the term "epidemic." But since the mid-1970s when the annual number of deaths from that disease reached a plateau, the term "epidemic" has not been even remotely applicable. Moreover, when measured by other indicators of mortality, the risk of dying from heart disease has declined steadily for the past fifty years. Indeed, the sustained downward trend in several measures of cardiovascular mortality is widely recognized within the community of cardiovascular specialists on the basis of numerous studies. That the accelerating decline in cardiovascular mortality was driven largely by a reversal in the trend for coronary heart disease (a major component of cardiovascular-renal diseases) was fully documented, intensively analyzed, and officially recognized in the national Conference on the Decline in Coronary Heart Disease Mortality held in 1978.[5] The Conference documented the inflection point at the end of the 1960s, when coronary heart disease reversed its long-standing upward course and began the decline that gave a major impetus to declines in cardiovascular disease mortality, which continues to the present.[6] Earlier analyses demonstrated that some components of cardiovascular-renal diseases, such as cerebrovascular diseases (stroke), hypertensive disease, and rheumatic fever and chronic rheumatic heart disease, began declining early in the twentieth century.[7] Today, the average risk of death from heart disease is about half that in 1950. While considerable progress has been made in preventing death from heart disease, the disease nevertheless remains an important health problem from both an individual and a public health point of view.

To provide insight into why there is still an argument about how to characterize the scope of heart disease in the United States, it is necessary to familiarize oneself with statistical indicators of health and death. Also, one must recognize that given an

[4] Rosenberg, "Cause of Death as a Contemporary Problem."

[5] Havlik and Feinleib.

[6] Rosenberg and Klebba.

[7] Moriyama, Krueger and Stamler.

array of alternative statistical indicators, advocacy groups are likely to select those that best support the message they wish to convey. Thus the AHA, described above, sought to characterize heart disease as a major and growing health menace in the United States by selecting statistical indicators that would communicate that message.

MEASURING MORTALITY

Presenting the scope and trend in diseases and deaths requires precision in both definition of the subject disease(s) and the statistical indicators. The subject disease is heart disease, defined using categories of the International Classification of Diseases (ICD, Ninth Revision) as diseases of heart including rheumatic fever and heart disease; hypertensive heart disease and renal disease; ischemic heart disease; other diseases of endocardium; and all other forms of heart disease.[8] The ICD-9 categories and codes were in effect for the years 1979-98. Corresponding codes for earlier years reflect the ICD in effect in the United States during those years; ICD-10 (the Tenth Revision of the ICD) is used to classify causes of death in U.S. mortality data effective with data for 1999.[9, 10]

Ischemic heart disease, often referred to as "coronary heart disease," accounts for about two-thirds of heart disease deaths, while "diseases of heart," which subsumes coronary heart disease, accounts for 70-80% of all deaths for the more inclusive category "cardiovascular disease." This analysis uses "diseases of heart," which is a better category than coronary heart disease for studying long-term trends because the latter exhibited major statistical discontinuities between successive revisions of the ICD. Such discontinuities make trend analyses difficult. Further, "diseases of heart" is a more inclusive category and is therefore less subject to variability in the terminology that physicians use to describe heart disease on death certificates.

Four statistical indicators of mortality are widely used in the presentation and analysis of mortality data. The indicators are as follows: numbers of deaths from heart disease; percent of deaths from heart disease; the crude death rate for heart

[8] World Health Organization, *Manual of the International Statistical Classification of Diseases, Injuries, and Causes of Death.*

[9] Revisions of the International Classification of Diseases (ICD) introduce discontinuities into time trends of causes of death, because of changes in categories and changes in coding rules. The degree of discontinuity is measured by "comparability ratios" which express the number of deaths in the later ICD divided by the number in the previous ICD. A ratio of 1.0 means that the net number of deaths for a cause of death in successive revision is exactly the same. Comparability for Diseases of heart between ICD-6 and ICD-7 was 1.0; between ICD-7 and ICDA-8, 1.0126; and between ICDA-8 and ICD-9, 0.9858. For a detailed discussion, see National Center for Health Statistics, *Vital Statistics of the United States.*

[10] The detailed classifications are as follows: Diseases of heart (ICD-9 Nos. 390-398, 402, 404-429) including Rheumatic fever and rheumatic heart disease (390-398), Hypertensive heart disease (402), Hypertensive heart and renal disease (404), Ischemic heart disease (410-414), Other diseases of endocardium (424), and All other forms of heart disease (415-423, 425-429). The ICD-9 categories and codes were in effect for the years 1979-98. Corresponding codes for earlier years reflect the ICD in effect in the United States during those years (ICD-9, 1979-1998; ICDA-8, 1968-78; ICD-7, 1958-67; ICD-6, 1949-57).

disease; and the age-adjusted death rate for heart disease. Each type of measure has particular strengths and weaknesses, and while each lends itself to a somewhat different interpretation, there is no ambiguity about which measure is technically appropriate when one wishes to present either the "burden of disease" or the "risk of death." These are two different concepts that call for different statistical indicators: *burden of disease* is reflected in the number, percent, and the crude death rate, while *risk of death* from the disease is reflected in the age-adjusted death rate.

The source of the data on mortality is death certificates filed in state vital statistics offices, and compiled into a national database by the National Center for Health Statistics.[11] Causes of death, such as heart disease, are reported on death certificates by physicians who attended the death; the demographic particulars of the decedent such as age, sex, and place of residence are reported by funeral directors based on information from informants, usually family members.

Indicator One: Number of Deaths from Heart Disease

In 1998, a total of 724,859 persons died of diseases of the heart.[12] Between 1950 and 1975, there was a sharp upward trend in the number of deaths from heart disease in the United States. The number of deaths in 1950 was 537,629; by 1973 the number grew to 757,075 (representing an increase of almost 10,000 per year). Since then, the numbers have more or less hovered between 700,000 and 800,000 per year. This measure indicates that, measured by numbers of deaths, the prevalence of heart disease has stabilized; thus, the present pattern of heart disease is not consistent with the characterization of an "epidemic."

The advantage of using the number of deaths as an indicator of mortality is its conceptual and computational simplicity. The disadvantage is that the number of deaths is sensitive to the size and demographic composition of the population; that is, the number of deaths will generally increase if the population increases (if other variables remain basically constant). Similarly, the number of deaths will decrease in the event that the population should decline. Further, if a segment of the population that is highly susceptible to heart disease increases in number (e.g. the aged), the number of deaths due to heart disease is likely to increase as well. In view of the gradual growth of the U.S. population (about 2-3 million per year), one would expect a gradual increase in the number of heart disease deaths, other things being constant.[13]

Indicator Two: Percent of Deaths from Heart Disease.

Another way of looking at the scope and trend of heart disease mortality is in relation to other causes of death (i.e. deaths from all causes combined), and expressing this relationship as a percent. This indicator is sometimes referred to as

[11] Murphy.
[12] Ibid.
[13] National Center for Health Statistics. *Health, United States, 2002,* Table 1.

"proportionate mortality," since it represents the proportion of all deaths accounted for by a particular cause of death.[14] Thus, in 1998, heart disease accounted for 31.9% of all deaths in the United States, or almost one out of three deaths.[15] The percent, or relative share, of mortality due to heart disease declined during the past 50 years. The percent stood at 37.0 in 1950, gradually increased to almost 39% in the mid-1980s, at about the same time as the number of deaths from heart disease reached its peak, but then declined sharply thereafter.

Like the number of deaths, the percent of deaths from heart disease is relatively simple to understand and calculate. The statistic tells us that in relation to other diseases, heart disease has diminished in relative importance over the past fifty years and that other diseases have become relatively more important. As an indicator of the scope and trend of heart disease mortality, however, the percent is seriously limited because it can increase or decrease over time regardless of whether deaths in the aggregate, or specific to heart disease, are increasing or decreasing. Thus, the percent of deaths due to heart disease in recent years has been decreasing while the annual number of deaths has been holding fairly steady.

Indicator Three: Crude Death Rate

The crude death rate is the number of deaths from heart disease in relationship to the size of the population in a given year. In 1998 a total of 724,859 persons died of heart disease, compared with a total population of the United States of 270,298,524.[16] The crude death rate is calculated by dividing the number of deaths in a year by the size of the mid-year population in that year, and then multiplying by a conventional "constant of proportionality" (usually 100,000) to make the number more intelligible. Thus, for 1998, the crude death rate is calculated as follows:

Crude death rate (CDR) = (annual number of deaths/mid-year population size) x 100,000

CDR = (724,859/270,298,524) x 100,000 = 0.002682 x 100,000 = 268 deaths per 100,000 population

In 1950, the crude death rate from heart disease in the United States stood at 356.8 deaths per 100,000 population.[17] The rate remained at about that level for two decades, until about 1970, after which it began decreasing sharply and almost steadily until by 1998 the rate was 25% below the 1950 level.

The crude death rate is a useful measure for a number of reasons. It permits one to quantify the "burden of mortality" from heart disease, that is, the extent of the burden that heart mortality imposes on the nation in relation to its population size. For the U.S. in 1998, the burden of heart disease mortality was about 2 to 3 deaths for every 1,000 population. Another advantage of the crude death rate is statistical:

[14] Moriyama.
[15] Murphy.
[16] Ibid.
[17] Ibid.

as a *rate*, which resembles a statistical probability, it is amenable to more sophisticated statistical analysis than are simple numbers or percents.

While the crude death rate is an interesting measure and can be useful to show how the burden of disease changes over time, or is distributed geographically, it has limitations as an analytical measure of mortality. Its principal limitation is its sensitivity to the demographic composition of the population. In other words, this statistic is sensitive to whether the population is relatively old or relatively young, or whether it is in the demographic process of aging (getting progressively older as time goes by), or getting younger. Thus, a crude death rate may increase simply because the elderly, who have a higher rate of dying from heart disease than the younger population, are becoming an increasing proportion of the population in the U.S. This is a serious limitation.

Indicator Four: Age-Adjusted Death Rate

A better measure of the risk of death is one that eliminates the distorting influence of the age composition of the population. Such an indicator is the age-adjusted death rate, also known as the age-standardized death rate.[18] The structure and properties of the age-adjusted death rate are similar to those of the crude death rate, but its internal components are weighted differently. Specifically, in a *crude death rate*, the death rate of each age component age group is, in effect, weighted by the actual size of the population in the corresponding age group. In the *age-adjusted death rate* the death rate of each age group is weighted by a "standard" population for that age group, where the standard population is arbitrarily held constant over time and among groups. For mortality statistics in the U.S., the population standard had been the estimated population of the U.S. for 1940; however, beginning with mortality data for 1999, the standard was changed to the projected population of 2000 and is used in this analysis.[19] The "standardization" process eliminates contamination from differences in age composition when comparisons are made of mortality risk over time or among groups. Specific procedures for calculating the age-adjusted death rate are described elsewhere.[20]

In 1998, the age-adjusted death rate for heart disease was 272.4 deaths per 100,000 standard population.[21] The age-adjusted death rate for heart disease declined during the past fifty years—initially somewhat slowly, that is, about half a percent per year from 1950-1965, and then much more rapidly, at a rate of about 1% decline per year through 1990. Recently, the rate of decline slowed to between 0.5-1.0% per year.

As a consequence of these sustained reductions, the age-adjusted death rate for heart disease mortality in 1998 is about half of what it was fifty years ago. This impressive decline in the death rate for heart disease, the leading cause of death in

[18] National Heart, Lung, and Blood Institute, *Comparing Death Rates over Time* and *Taking the Mystery Out of the Age-Adjusted Death Rate;* Murphy.

[19] Anderson and Rosenberg.

[20] Ibid.

[21] National Center for Health Statistics, *Historic data for death rates.*

the U.S., is a major contributor to the 8.5 years added to the average length of life in the U.S. during 1950-1998.[22] The trend in the age-adjusted death rate shows that heart disease is not technically an "epidemic" in the United States.

The advantage of the age-adjusted death rate as a comparative indicator of mortality is that it takes into account the impact of age distribution. One disadvantage of the measure is that the age-adjusted death is a theoretical construct, or a hypothetical index, because an integral element of the measure is an arbitrary set of weights (called the "standard population"). Another disadvantage is the conceptual complexity of the measure, which sometimes militates against its use in effectively communicating a public health message. Nonetheless, the analytical benefits of being able to compare mortality risks among groups or over time far outweigh the limitations of age-adjusted death rate for assessing trends and patterns of mortality.[23]

Comparative Analysis

The four indicators of mortality can be compared directly (Figure 1) by expressing each as an index with a value of 1.0 for the calendar year 1950 used as a benchmark. The index numbers for each measure depict trends similar to those listed in previous sections prior to indexing. By indexing, the scale for all the measures is comparable, and their starting point in 1950 is 1.0. One can examine the trend for each indicator. Thus, for the number of deaths, the index for the most recent year, 1998, is 1.35, which means that the number of deaths from heart disease in 1998 is 35% above the number reported in 1950. For the measure, percent of deaths due to heart disease, the index for 1998 is 0.84, indicating that compared to all causes of death the percent of deaths due to heart disease declined by about 16% between 1950 and 1998. During the same period, the index for crude death rate changed from 1.0 to 0.75, representing a decline of 25%. For the age-adjusted death rate, the index at 1998 stood at 0.46, representing a decline in heart disease mortality of 54%, or over half.

[22] Murphy.
[23] Anderson and Rosenberg.

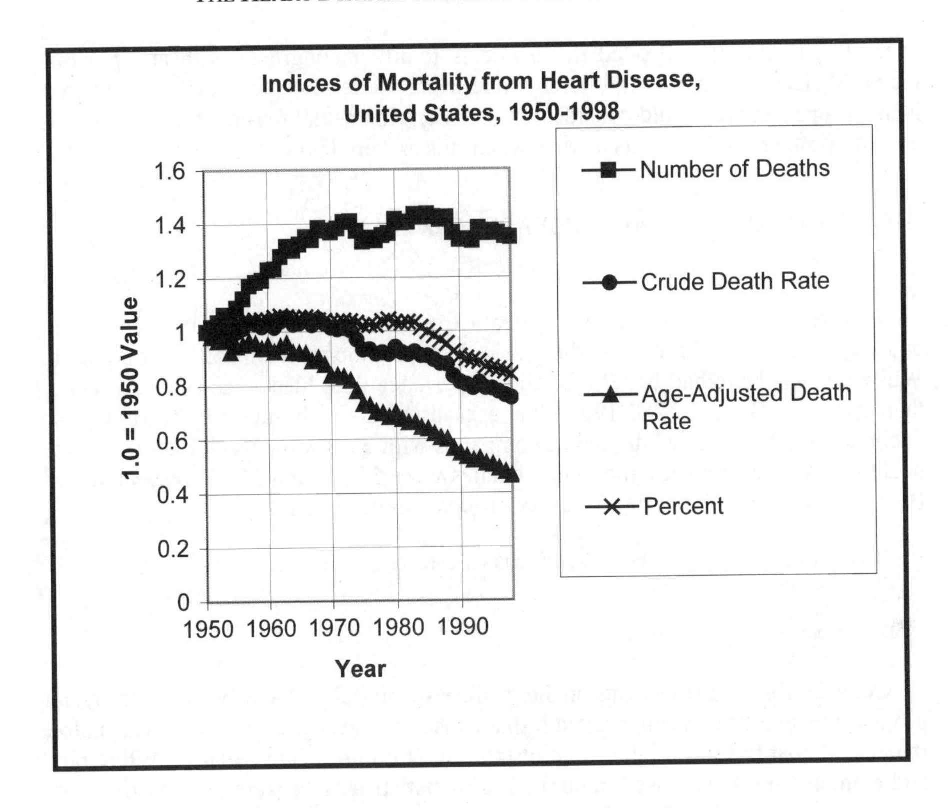

Figure 1. Indices of Mortality from Heart Disease.

The relative decline of the age-adjusted death rate is almost twice that of the crude death rate, showing that *the aging of the U.S. population masks the substantial reduction in the risk of mortality from heart disease*. The reduction of over 50% in the relative risk of heart disease mortality—reflected in the age-adjusted death rate—is twice as great as the reduction of 25% in the burden of death from heart disease—reflected in the crude death rate.

Thus, these statistical analyses challenge the conventional wisdom that the U.S. is experiencing an epidemic of heart disease mortality. While increases in the number of deaths due to heart disease from 1950-73, averaging about 10,000 per year, might have been characterized by some as epidemic increases, a generation later such a characterization is not consistent with the facts. Even the most simplistic measures of mortality—the annual number of deaths from heart disease—has fluctuated within a relatively narrow range of 700,000–800,000, and recently closer to the lower limit of this range. Trends for all the other indicators of heart disease mortality are strongly at odds with the epidemic characterization; that is, they show greater and greater declines in heart disease mortality as the indicators are of a progressively refined nature. A reduction in the risk of heart disease mortality of over 50% in the age-adjusted death rate is totally incongruent with an epidemic increase! If we as a society are to continue using the term "epidemic," even when it is inappropriate, we should consider some underlying social reasons for doing so. Is a health problem only seen as a crisis when it is an "epidemic"?

DIFFERENTIALS IN HEART DISEASE MORTALITY

The steady decline in the risk of death from heart disease has not been shared equally by all race, ethnic, and other socio-demographic segments of the population. White persons have had better mortality experience than blacks, and females better than males.[24] During 1950-1998, the age-adjusted death rate for heart disease declined by 52.5% for white males, compared with 37.6% for black males; and the declines for white females and black females were 54.5% and 46.7% respectively.[25] By 1998, rates among the four race-sex groups ranked as follows:

White females	217.6 deaths per 100,000 standard population
Black females	291.9
White males	332.2
Black males	407.8

Considerable improvements in heart disease mortality have been made by all groups; however, the death rate from this cause for the highest group, black males, remains almost twice as high as that of the lowest group, white females. While race and ethnicity are widely used in public health statistics as variables of classification, it is well understood that the differentials they reflect are largely attributable not to race/ethnicity *per se*, but rather to social, educational, and economic factors that characterize these groups. Further, imprecision in reporting race and ethnicity on administrative records such as death certificates, and in censuses and surveys has been widely documented.[26] In assessing trends and differentials in mortality by race and ethnicity, it is therefore important to understand that they represent, though imperfectly, the effect on health of socio-demographic characteristics of the population. Further, it is increasingly recognized that in American society many people do not fit neatly into one race or ethnic category. Thus, the year 2000 Census of Population asked respondents to check the multiple race/ethnic categories to which they belonged rather than asking them to identify with a single group.

How social, economic, and health statistics will accommodate the multi-racial background of many Americans is an important question, both ethically and statistically. The degree to which our statistical categories accurately represent the population has consequences for public health and government programs, and the

[24] "White" and "Black" are among the standard categories specified by the U.S. Office of Management and Budget for classification of race in statistical data. See U.S. Office of Management and Budget, "Standards for Maintaining, Collecting, and Presenting Federal Data on Race and Ethnicity." See also the internet reference, http://www.doi.gov/diversity/doc/racedata.htm.

[25] National Center for Health Statistics, *Historic data for death rates.*

[26] Rosenberg et al, "Quality of death rates by race and Hispanic origin."

importance (and difficulty) of determining these categories should not be underestimated.

A well-known and widely documented determinant of health and mortality is educational attainment, which also accounts, in part, for differentials in mortality by race.[27] Educational attainment influences health in a number of ways: increasing educational attainment is associated with more knowledge about prevention, healthier life styles, and knowledge about treatment; and educational attainment is, to a large degree, a determinant of higher income, which provides better access to the health care system.

Wide disparities in heart disease mortality are associated with educational attainment. They can be measured using a "standardized mortality ratio" (SMR). SMRs show the *observed* mortality of a selected group compared with the *expected* mortality of a benchmark group, expressed as a ratio (times 100). The benchmark group is usually the total population. Thus, we can compare the mortality of persons with different levels of education with the mortality of the total population of all educational attainment levels. An SMR of 100 means that the selected group exhibits mortality the same as the general population; an SMR of over 100 indicates higher-than-average mortality, while an SMR of less than 100, lower-than-average mortality. Large differentials in mortality by educational attainment were shown in the classic study of socio-economic differentials in mortality by Kitagawa and Hauser, based on the 1960 Census of Population and death records.[28]

A more recent data base for deaths occurring during 1979-85 shows that educational attainment continues to be one of the most powerful determinants of mortality, including that from heart disease.[29] Thus, for white males, twenty-five years old and over who completed eight years of schooling, the SMR was 112 or 12% above the average of white males of all educational attainments combined. The ratio of white males who completed four years of college was 75, indicating mortality 25% *below* the average of 100. Thus, mortality for the most highly educated group of white males was 37% lower (12+25) than for the least educated, reflecting the powerful association between educational attainment and subsequent mortality.

The table below indicates that educational attainment is a powerful determinant of heart disease mortality for each of the race-sex groups, with the greatest educational differences evident for black males, ranging from 21% above the average mortality for those with eight years of schooling to 43% below average (100-57) for those who completed college. It is particularly striking that black males have both the lowest *and* highest mortality due to heart disease when correlated with education. Because of the apparent link between education and mortality, it is clear that provision of educational opportunities for all members of society can have substantial public health benefits.

[27] National Center for Health Statistics, "Chartbook on socioeconomic status and health."
[28] Kitagawa and Hauser.
[29] Rogot et al., Table 6.

Educational Attainment	White Males	White Females	Black Males	Black Females
8 years	112	109	121	100
12 years	98	105	83	90
College	75	70	57	66

MORBIDITY

The health of a population is often described using mortality indicators because of their quality, availability, and relative uniformity over time and among areas.[30] However, a comprehensive view of health requires that mortality (death) data be complemented with morbidity (illness) data such as statistics on the prevalence on heart disease and associated risk factors such as serum cholesterol level and hypertension.

The prevalence of heart disease in the U.S. is measured by the National Center for Health Statistics using health examinations for a representative national sample of households.[31] Between 1976-80 and 1991-94, the prevalence of heart disease for persons aged 25-74 years declined for all males and all females, and for the white population (Figure 2).[32] For the black population, prevalence increased during this period.[33] Age-adjusted prevalence in 1991-94 for the white population was 4.75%, while that for the black population was 6.97% or 41% percent higher than for the white population. The prevalence for males was 18% higher than for females during that period, 5.53% and 4.69%, respectively.

Changes in the prevalence of heart disease during this period were accompanied by sharp reductions in serum cholesterol levels and hypertension. The age-adjusted percent of the population aged 25-74 years with high serum cholesterol declined from 33.3% in 1960-62 to 19.7% in 1988-94, a reduction of over one-third.[34] Similarly, the age-adjusted percent of the population with hypertension declined during this period from 38.1% to 23.9%—a reduction of 37%.[35]

[30] Rosenberg, "Improving cause-of-death statistics," 563-564.
[31] National Center for Health Statistics, *National Health and Nutrition Examination Survey*, 360-361.
[32] Percents are age-adjusted to the year 2000 standard population.
[33] National Heart, Lung, and Blood Institute, *Morbidity and Mortality*.
[34] National Center for Health Statistics, *Health, United States, 2002*, Table 69.
[35] Ibid., Table 68.

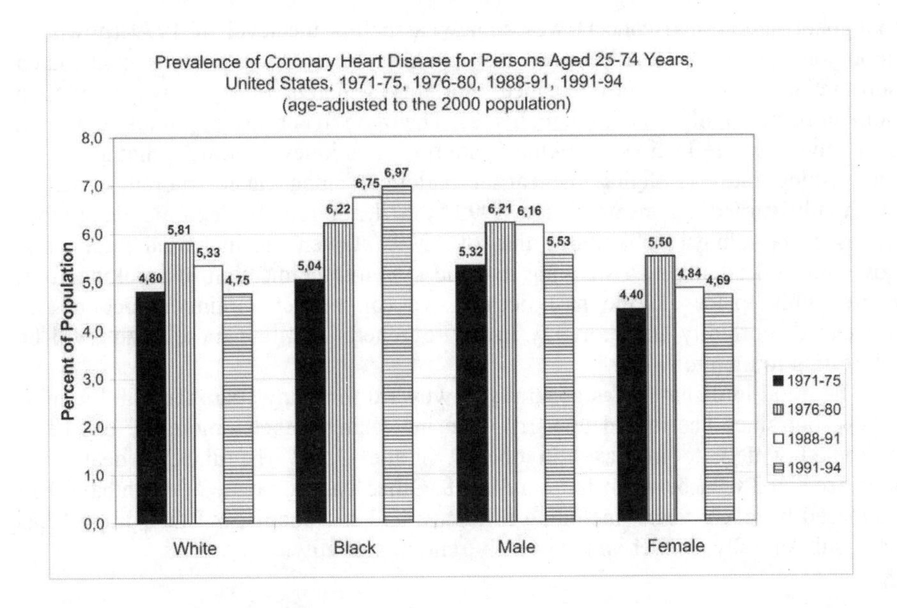

Figure 2. Prevalence of Coronary Heart Disease, 1950-1998.

OTHER CHANGES IN MORTALITY

It is clear that the "epidemic" characterization of heart disease in the U.S. is not appropriate. Nonetheless, heart disease is still the leading cause of death in the United States, accounting for almost one-third of all deaths. Thus, heart disease mortality strongly influences overall mortality. In addition, the substantial improvements in life expectancy in the United States during the past fifty years are, to a large extent, the result of improvements in heart disease mortality. Despite overall improvements in mortality and life expectancy, however, there have been some notable countervailing forces in recent years, such as mortality from cancer (the second leading cause of death in the United States); chronic respiratory disease (the fourth leading cause); and diabetes (the seventh leading cause) (Table 1, at Appendix A).[36]

For cancer, mortality increased steadily through 1990, when it began a decline that continues through the present.[37] By 1998, the age-adjusted death rate for

[36] Murphy.

[37] Edwards et al., "Annual report to the Nation on the status of cancer."

malignant neoplasms (cancer) was somewhat below the level of 1979 (down by about one percent). It is believed that changes in smoking patterns contributed substantially to the downturn in cancer mortality, which is reflected in declines for a number of major sites, though not all sites. The trend in chronic respiratory diseases, probably also reflecting smoking patterns, continues upward, though at a diminishing rate paralleling the earlier pattern of lung cancer mortality, which eventually turned downward. By 1998, the death rate for chronic respiratory diseases was almost 65% above that of 1979. Given the increased public and governmental attention to smoking and the systemic eradication of smoking from many public places in the past decade, we can expect continuing declines in associated morbidity and mortality in future generations, and these trends will be interesting to monitor.

The trend in diabetes was generally downward for many years through the mid-1980s, but since has turned upward. The increasing diabetes mortality has been attributed widely to increases in obesity in the U.S.[38] The diabetes death rate increased by 38.3% between 1979 and 1998. Other leading causes of death have also increased in recent years, including influenza and pneumonia, but their impact has been substantially smaller on the overall trend in mortality.

REVISITING THE POLITICS OF HEALTH INDICATORS

As evidenced by the four indicators of heart disease mortality, different measures can tell different stories ranging, in the case of heart disease, from increases (numbers of deaths during 1950 through 1975, followed by stabilization) to precipitous declines (age-adjusted death rates). As noted in the introduction, indicators can sometimes be selected to support a particular case. But as is evident from the above discussion, one or two indicators do not tell the whole story; each indicator has particular strengths and weaknesses, and the best assessments of health trends occur when the evidence from all possible indicators are considered together. The example of the American Heart Association versus National Heart, Lung and Blood Institute is but one example of this.

In addition, there are ethical questions pertaining to the methodology of health statistics. For example, in an increasingly multiracial and multicultural society, how are health statisticians to classify persons of mixed race and ethnic heritage? What are the societal implications of the ways in which we choose to assign racial background? In terms of the social effects of health statistics, we might ask what society's responsibility is to underprivileged persons who do not have easy access to higher education, or for whom financial needs preclude a college education. Since overall health and heart disease in particular strongly correlate with level of education, it seems important to stress higher education as a public health initiative in its own right! The correlations articulated by health statistics can thus influence social programs and, hopefully, make public policy accountable to the well-being of the people that it serves.

[38] Taubes.

CONCLUSION

In his article in *Science* magazine, Taubes linked the dietary campaigns of the 1990s, which continue unabated today, with the commonly held view that heart disease in the U.S. is increasing rapidly, indeed at epidemic proportions. Such a view is consistent with the interests of groups representing the food and diet industries and the organizations that advocate for the heart disease community. Mortality data from the National Center for Health Statistics, however generously interpreted, are at variance with this perception. The least sensitive indicator of heart disease mortality—numbers of deaths—has been essentially stable for about thirty years while the most sensitive indicator—age-adjusted death rates—has declined substantially during the past fifty years.

One can conjecture, conservatively and with reasonable confidence, about the prospects for heart disease mortality in the United States. One needs to distinguish between future *rates* and future *numbers* of deaths. In terms of rates (both crude and age-adjusted) we can expect continued declines, because of improvements in prevention and treatment, as well as changes in the dynamics affecting the social profile of the population, particularly improvements in educational attainment. Education, we have demonstrated, is one of the most powerful correlates of heart disease mortality, and educational attainment of the U.S. population continues to improve. Thus, the percent of the population who graduated from college increased from 7.7% of the population in 1960 to 24.4% in 1998,[39] and shows no evidence of a plateau. As the population becomes increasingly well-educated, the health behaviors and characteristics of the better educated population can be expected to widely diffuse throughout the population affecting not only heart disease but other causes of death. For example, as a result of the increasingly higher educational attainment of the population it can be expected that a larger share of the population will experience the much lower risk of heart disease mortality now prevailing among those who are college graduates. In terms of future numbers of deaths, two countervailing forces are operative: on the one hand, death rates are declining, which disposes toward smaller number of deaths; on the other hand, the population is growing (largely because of immigration) and aging, which results in larger numbers of deaths. These two opposing forces have been largely in balance for the past thirty years, resulting in a relatively stable number of annual deaths from heart disease from 1973 to the present. This is likely to be the situation in the immediate and long-term future. More definitive projections require statistical modeling using explicit assumptions about future mortality, fertility, and migration. However, statistical modeling is *not* required to assert that the U.S. is unlikely to ever experience increases in numbers of heart disease deaths similar to the epidemic that occurred through the late 1960s. The momentous downturn in coronary heart disease mortality in the late 1960s is the statistical and epidemiological foundation for reduced heart disease prevalence now and in the future, barring unforeseen circumstances.

[39] U.S. Bureau of the Census.

While the prevalence of heart disease is, indeed, substantial—afflicting about five in 100 persons—it shows signs of declining for the white population, though not for the black population. For both blacks and whites, two important indicators of heart disease risk—serum cholesterol level and hypertension—have declined substantially during the past thirty years, contributing to the decline in heart disease mortality.[40] Clearly, then, progress is being made in the prevention, diagnosis, and treatment of heart disease. Advocacy groups deserve credit for these gains by raising public awareness and by exerting influence on the government to provide resources for research and education. To the extent that their efforts are based on an objective portrayal of the health problem, their work deserves full support, recognition, and appreciation. However, to the extent that their policies are based on a deliberately skewed and misleading presentation of statistics, they should be exposed and challenged.

Health statistics collected by state and federal agencies are the empirical foundation upon which policy makers must evaluate what health problems to address, and monitor progress in addressing these problems. Health statisticians have an obligation to not only demonstrate which statistical indicators are most appropriate to characterize the scope and tempo of the problem, but also to identify unscrupulous and technically flawed uses of the data, no matter how worthy the program goals in whose service the statistics are being used.

[40] See notes 33 and 34.

APPENDIX A:

Recent Trends in the Leading Cause of Death in the United States, 1979-1998

TABLE 1. RECENT TRENDS IN THE LEADING CAUSES OF DEATH IN THE UNITED STATES, 1979-1998

[Death rates are on an annual basis per 100,000 population. Age-adjusted death rates are standardized to the year 2000 population, and are expressed per 100,000 standard population.]

Rank	Cause of Death (ICD-9)	Percent of total deaths 1998	Crude death rate 1998	Age-adjusted death rate 1998	Percent change in age-adjusted death rate 1979-98
	(1)	(2)	(3)	(4)	(5)
...	All causes	100.0	864.7	875.8	-13.4
1	Diseases of heart	31.0	268.2	272.4	-32.2
2	Malignant neoplasms	23.2	200.3	202.4	-0.8
3	Cerebrovascular diseases	6.8	58.6	59.6	-38.7
4	Chronic obstructive pulmonary diseases	4.8	41.7	42.0	64.7
5	Accidents and adverse effects	4.2	36.2	36.3	-24.2
...	Motor vehicle accidents	1.9	16.1	16.1	-28.6
...	All other accidents and adverse effects	2.3	20.1	20.2	-20.2
6	Pneumonia and influenza	3.9	34.0	34.6	32.6
7	Diabetes mellitus	2.8	24.0	24.2	38.3
8	Suicide	1.3	11.3	11.3	-10.3
9	Nephritis, nephrotic syndrome, and nephrosis	1.1	9.7	9.8	14.0
10	Chronic liver disease and cirrhosis	1.1	9.3	9.5	-35.8
11	Septicemia	1.0	8.8	8.9	106.9
12	Alzheimer's disease	1.0	8.4	8.6	4200.0
13	Homicide and legal intervention	0.8	6.8	6.7	-32.3
14	Atherosclerosis	0.7	5.7	5.8	-67.6
15	Hypertension with or without renal disease	0.6	5.3	5.4	31.7
...	All other causes	15.8	136.6	...	...

... Category not applicable

Note: Rank is based on number of deaths.

Sources: Cols. (2) and (3) from Murphy SL. Deaths: final data for 1998. National Vital Statistics Reports. Vol. 48. No. 11. Hyattsville, Maryland. National Center for Health Statistics; cols. (4) and (5) from unpublished historic data at http://www.cdc.gov/nchs/datawh/statab/unpubd/mortabs/hist293.htm

REFERENCES

Anderson, R.N. and H.M. Rosenberg. "Age standardization of death rates: Implementation of the year 2000 standard." *National Vital Statistics Reports* 47, no. 3. Hyattsville, Maryland: National Center for Health Statistics, 1998.

Bishop, J.E. "Heart disease may actually be rising." *Wall Street Journal*, Eastern Edition 228, no. 96 (November 13, 1996): B-6 (2 graphs).

Edwards, B.K., H.L. Howe, L.A.G. Ries, M.J. Thun, H.M. Rosenberg, R. Yancik, P.A. Wingo, A. Jemal, E.G. Feigal. "Annual report to the Nation on the status of cancer, 1973-1999, featuring implications of age and aging on U.S. cancer burden." *Cancer* 94, no. 10 (2002): 2766-2792.

Havlik, R.J. and M. Feinleib. *Proceedings of the Conference on the Decline in Coronary Heart Disease Mortality*. National Heart, Lung, and Blood Institute, National Institutes of Health, October 24-25, 1978. National Institutes of Health, NIH Publication No. 79-1610. May 1979.

Kitagawa, E.M. and P.M. Hauser. *Differential Mortality in the United States: A Study in Socioeconomic Epidemiology*. Cambridge: Harvard University Press, 1973.

Moriyama, I.M., D.E. Krueger and J. Stamler J. *Cardiovascular Diseases in the United States*. Cambridge: Harvard University Press, 1971.

Murphy, S. "Deaths: final data for 1998." *National Vital Statistics Reports* 48, no. 11 (July 24, 2000). Hyattsville, Maryland: National Center for Health Statistics, 2000.

National Center for Health Statistics. "Chartbook on socioeconomic status and health." *Health, United States, 1998*. U.S. Department of Health and Human Services, Centers for Disease Control. DHHS Publication No. 1232.

________. *National Health and Nutrition Examination Survey*. Health United States, U.S. Department of Health and Human Services, Centers for Disease Control. DHHS Publication No. 1232. August 2002: 360-361.

________. *Health, United States, 2002*. U.S. Department of Health and Human Services, Centers for Disease Control. DHHS Publication No. 1232. August 2002.

________. *Historic data for death rates age-adjusted to the year 2000 standard population.* 2003. http://www.cdc.gov/nchs/datawh/statab/unpubd/mortabs/hist293.htm

________. *Vital Statistics of the United States,* Vol. I, "Mortality," Technical Appendix. Hyattsville, Maryland: National Center for Health Statistics.

National Heart, Lung and Blood Institute. *Comparing Death Rates Over Time: Data Fact Sheet.* National Institutes of Health, September 1997.

________. *Morbidity and Mortality: 2000 Chart Book on Cardiovascular, Lung, and Blood Diseases.* National Institutes of Health. May 2000. Chart 3-16, based on unpublished tabulations from the pubic use data tapes, National Health and Nutrition Examination Survey. National Center for Health Statistics, 1999.

________. *Taking the Mystery Out of the Age-Adjusted Death Rate: Data Fact Sheet.* National Institutes of Health, September 1997.

Rogot, E., P.D. Sorlie, N.J. Johnson, and C. Schmitt. "A Mortality Study of 1.3 Million Persons by Demographic, Social, and Economic Factors: 1979-85 Follow-Up: U.S. National Longitudinal

Mortality Study." National Institutes of Health, National Heart, Lung, and Blood Institute, NIH Publication No. 92-3297. July 1992. Table 6.

Rosenberg, H.M. "Cause of Death as a Contemporary Problem." *Journal of the History of Medicine and Allied Sciences* 54 (April 1999): 133-153.

________. "Improving cause-of-death statistics." *American Journal of Public Health* 79, no. 5 (May 1989): 563-564.

Rosenberg, H.M. and A.J. Klebba. "Trends in cardiovascular mortality with focus on Ischemic heart disease: United States, 1950-1976." In *Proceedings of the Conference on the Decline in Coronary Heart Disease Mortality.* Edited by Havlik, R.J. and M. Feinleib. National Heart, Lung, and Blood Institute, National Institutes of Health, October 24-25, 1978. National Institutes of Health, NIH Publication No. 79-1610. May 1979.

Rosenberg, H.M., J.D. Maurer, P.D. Sorlie, N.J. Johnson, M.F. MacDorman, D.L. Hoyert, J.F. Spitler, and C. Scott. "Quality of death rates by race and Hispanic origin: a summary of current research, 1999." *Vital and Health Statistics* 2, no. 128. Hyattsville, Maryland: National Center for Health Statistics, September 1999.

Taubes, G. "The soft science of dietary fat." *Science* 291 (2001): 2536-2545.

U.S. Bureau of the Census. *Statistical Abstract of the United States, 2001*. Suitland, Maryland, 2002.

U.S. Office of Management and Budget. "Standards for Maintaining, Collecting, and Presenting Federal Data on Race and Ethnicity." *Federal Register*. October 30, 1997: Appendix A. Information can also be found on the internet, http://www.doi.gov/diversity/doc/racedata.htm.

World Health Organization. *Manual of the International Statistical Classification of Diseases, Injuries, and Causes of Death: International Classification of Diseases* (Based on Recommendations of the Ninth Revision Conference, 1975, and Adopted by the Twenty-ninth World Health Assembly). Geneva: World Health Organization, 1977.

JAMES FLORY & EZEKIEL EMANUEL

RECENT HISTORY OF END-OF-LIFE CARE AND IMPLICATIONS FOR THE FUTURE

INTRODUCTION

End of life is a critical and changing area of health care. In the mid 1970s, quality of dying was brought to the widespread attention of the American public through the introduction of hospice to the United States, in 1974,[1] and especially by the *Quinlan* court case, in 1976.[2] The image of technology trapping a twenty-one year old woman in a vegetative half-life spurred the "right to die" debate and the living will movement. Just months after the case, California became the first state to enact a living will law. Judicial decisions and legislation continued through the 1980s, including more right-to-die cases at the state level,[3] the introduction of the Medicare hospice benefit, and the Patient Self-Determination Act, which passed Congress in 1991. In the 1990s concern about the right to die produced three Supreme Court cases[4] and extensive popular attention.[5] The decade also saw energetic attempts to improve the quality and reduce the high cost of caring for the dying. Examples include rapid growth in the use of palliative and hospice care,[6] funding for improvements to end of life care from private foundations,[7] and major studies on end of life.[8]

The ideal for end-of-life care is to give dying patients comfort, dignity, and a "good death." Polls and studies have shown that the majority of Americans prefer to die at home if possible.[9] The popular imagination idealizes death in one's own bed, surrounded by friends and family. At the other extreme, many people dread death in an impersonal hospital ward, bound by multiple intubations and other high technology invasions. The medical humanities have done a particularly good job of

[1] Field and Cassel, eds.
[2] *In Re Quinlan* (1976).
[3] Cantor, *Legal Frontiers of Death and Dying.* Major cases from the 1980s cited include *Lydia E. Hall Hospital, Conroy*, and *Bartling.*
[4] *Cruzan by Cruzan v. Director* (1990); *Washington v. Glucksberg* (1997); *Vacco, Attorney General of New York v. Quill* (1997).
[5] E.J. Emanuel, "Why Now?"
[6] Field and Cassel, eds.
[7] e.g., Robert Wood Johnson Foundation, the Nathan Cummings Foundation, and the Commonwealth Fund
[8] e.g., Commonwealth-Cummings and the SUPPORT study (SUPPORT Principal Investigators).
[9] SUPPORT Principal Investigators; Gallup Organization; Pritchard et al.

A. W. Galston and C. Z. Peppard (eds.), Expanding Horizons in Bioethics, 161-182.

finding powerful expressions of these feelings. From medieval woodcuts to neoclassical idealizations of the death of Socrates, we can find images that make death almost attractive by depicting comfort, dignity, and a degree of choice (Figure 1). In contrast, some modern works show the nightmare that invasive life-sustaining care creates in the modern mind.[10] Though it would be wrong to care for all dying people according to one rigid model, none should object to the system's making a serene and meaningful death possible for as many terminally ill patients as can have it and want it.

Even though the ideal of a "good death" creates a meaningful goal with broad appeal, it leaves many ethical and policy questions unanswered. Different parts of the population have different needs at the end of life and different levels of access to palliative care. Meeting diverse needs and solving problems of unequal access is made all the more challenging because the way dying patients are cared for has been changing for decades, under influences that are not well understood. This chapter is an attempt to introduce the most pressing of these policy and ethical issues, in the context of the major trends and innovations in end-of-life care over the last thirty years of the twentieth century. We will focus on a set of issues that seem certain to affect very large numbers of people. This means that we will not address some very rich issues—like euthanasia—that are of less sweeping importance from a policy standpoint.

The setting of death—in a hospital's inpatient bed, at home, in a nursing home, or elsewhere—is a relatively well-tracked national indicator that provides some insight into what end-of-life care in America is like. The first section of this chapter will summarize what changes in place of death can tell us about how care for the dying evolved at the end of the twentieth century. Second, we will discuss changes in policy and practice during the same period, and third, we will discuss the ways in which demographic factors appear to affect end-of-life care. These considerations will lead to a brief discussion of some key ethical and policy questions. What kinds of research are most important for improving the quality of end-of-life care? What kinds of end-of-life care do not currently receive adequate oversight? Does everybody in America have access to the same quality of end-of-life care? Do disparities need to be addressed through policy? How do we balance needs for improving end-of-life care against the financial cost of doing so? Finally, what are going to be the key issues in end-of-life care in the near future?

[10] Bertman.

Figure 1. "The Goodman on His Deathbed" (1471), Anonymous. Woodcut. Illustration from German Bibliotheque des Artes Decoratifs, Paris. Blockboard Edition of *Ars Moriendi,* Bibliotheque des Artes Decoratifs, Paris. Copyright © 1991 from *Facing Death: Images, Insights and Interventions* by Sandra Bertman. Reproduced by permission of Routledge/ Taylor and Francis Books, Inc.

OVERVIEW OF DEATH IN AMERICA

In 2000, just over 2.4 million people died in America. The leading overall causes of death were heart disease, cancer, and stroke (Table 1). Heart disease and cancer alone accounted for just over half of all deaths. HIV has dropped out of the top 15 causes of death for Americans; in 2000 it killed roughly 14,000 people, fewer than died of homicide.

Mortality patterns are different for different races and ethnic groups. The importance of cancer and heart disease is universal, but for Hispanics and American Indians, accidents displaced stroke as the number three cause of death in 2000. A disturbing feature of the racial and ethnic breakdown was that homicide ranked as the sixth and seventh cause of death among African-Americans and Hispanics, respectively.[11]

Decedents tend to be older: people over the age of 65 accounted for almost 75% of all deaths in the United States.[12] The overwhelming majority of those older decedents were covered by Medicare.[13] Because of Medicare, most decedents are relatively well-insured, and this also means that Medicare policies have a powerful effect on end-of-life care. The significance of Medicare to end-of-life care will only increase in the coming decade, as the percentage of Americans above the age of 65 rises.

Table 1. Top Ten Causes of Death in the United States. Reprinted with slight modifications from R.N. Anderson and B.L. Smith, "Deaths: Leading Causes for 2001." *National Vital Statistics Reports* 52 (9), 2003.

Cause	*Number of Deaths*	*Percentage of Total*
Heart Disease	700142	29.0
Cancer	553768	22.9
Stroke	163538	6.8
Chronic Lower Respiratory Disease	123013	5.1
Accidents	101537	4.2
Diabetes	71372	3.0
Influenza and Pneumonia	62034	2.6
Alzheimer's Disease	53852	2.2
Kidney Disease	39480	1.6
Septicemia	32238	1.3

[11] Minino et al.

[12] Calculated from Minino et al.

[13] Field and Cassel, eds.

CHANGES IN PLACE OF DEATH

End-of-life care in America has been changing for at least the past half-century. One way of seeing this is to look at where people have been dying. The late seventies and early eighties appear to mark the peak of a long increase in the proportion of deaths occurring in the hospital as medicine in the United States modernized. But, since that time, the health system has evolved away from deaths in the hospital toward deaths at home and nursing homes.

For the last two decades, the rate of in-hospital death in the United States has been declining at a fairly constant rate of nearly one percent per year. Data prior to the 1990s are approximate, but indicate that over two decades the proportion of deaths occurring in-hospital has fallen about 13%, from a high of over 50% to about 40% (Figure 2). From 1990 to 1998 alone, when the data are more reliable, the decline has been 8%. The sustained decline of in-hospital deaths has quietly altered the experiences of millions. Approximately 310,000 people each year—all of whom would have died in the hospital if the proportion of in-hospital deaths had remained unchanged from the 1980s—currently die outside of the hospital setting.[14]

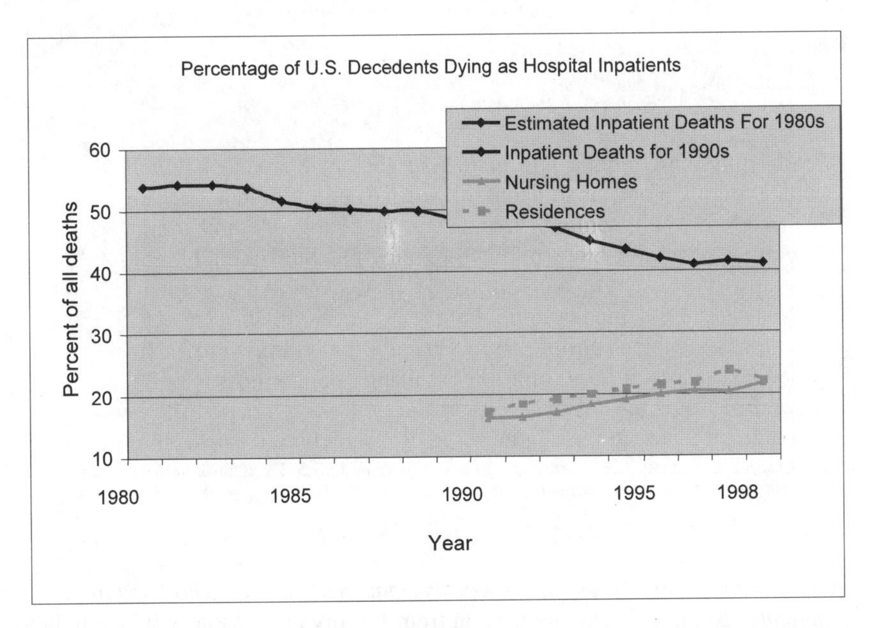

Figure 2. Percent of Decedents in the United States Dying as Hospital Inpatients. Reprinted with permission from J. Flory, Y. Young-Xu, I. Gurol, N. Levinsky, A. Ash, E. Emanuel, "Place of Death: U.S. Trends Since 1980," *Health Affairs* 23 (2) 2004: 194-200. Reprinted with permission of the publisher: *Health Affairs,* http://www.healthaffairs.com.

[14] Flory et al.

While in-hospital deaths have declined for every major cause of death, the decline has been greater for some causes than others (Figure 3). Cancer's rate of in-hospital death has declined more than that of any other disease. In the 18 years between 1980 and 1998, death from cancer went from occurring in a hospital bed about 70% of the time to just 37%. Cancer is responsible for half of the overall change in place of death, despite accounting for only a quarter of deaths. In-hospital deaths from chronic obstructive pulmonary disease declined by about 20 percentage points over the same time period. Other major causes also declined, but not as significantly: about fifteen percentage points for diabetes and chronic liver disease, about thirteen percentage points for heart disease and stroke, and less than ten percentage points for pneumonia and influenza. It appears that deaths with slow and predictable courses have moved out of the hospital more than those that strike acutely.

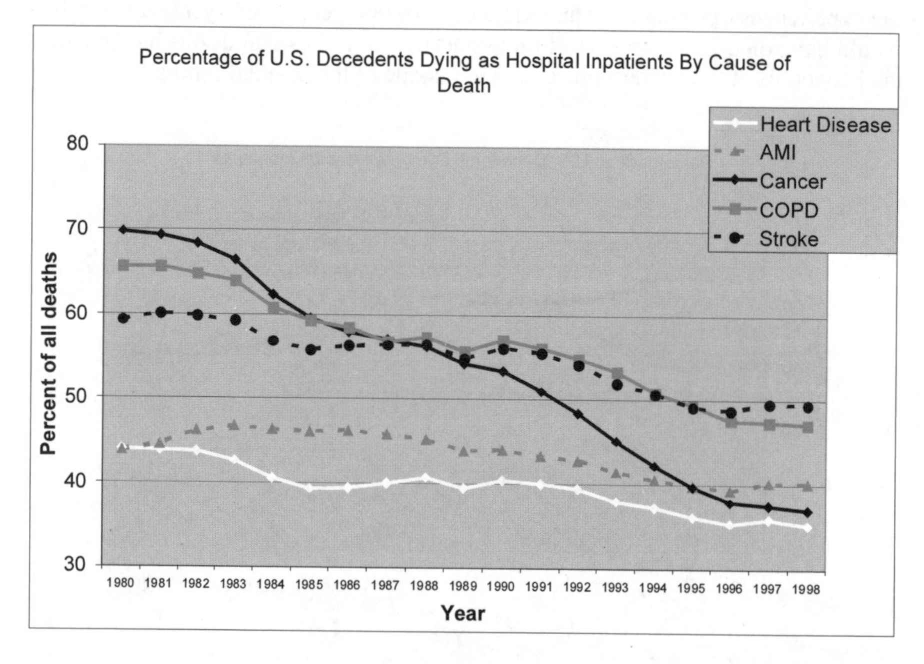

Figure 3. Changes in Percent of Decedents from each Disease who Died as Hospital Inpatients. Reprinted with permission from J. Flory et al., "Place of Death: U.S. Trends since 1980," *Health Affairs* 23 (2) 2004: 194-200. Reprinted with permission of the publisher: *Health Affairs,* http://www.healthaffairs.com.

CHANGES IN CARE

The broad decline in hospital deaths has accompanied, and has perhaps been partly caused, by several major changes in practice and policy. Innovations related to end-of-life care in the last three decades have included the use of advance directives, the rise of Medicare-funded hospice, increased use of nursing homes and at-home care, and Medicare's use of capitation.

Advance Directives and Advance Care Planning

The end of life often demands difficult choices, such as whether or not to use life-sustaining techniques like ventilators, cardiopulmonary resuscitation, or intravenous feeding. Patients in a position to receive such interventions are frequently incapable of making decisions, creating potentially excruciating legal and ethical problems for those who must decide instead. A famous 1990 Supreme Court decision, *Cruzan v. Director, Missouri Health Department*, addressed a case in which a woman named Nancy Cruzan was in a comatose state and was kept from death by intravenous feeding. After several years had passed, family members wanted life support withdrawn, but local authorities would not permit the withdrawal without court approval. The state of Missouri ruled that clear and convincing evidence of Cruzan's preference—such as written documentation or a verbal expression more serious than a casual comment—was needed before it could permit cessation of life support. The Supreme Court supported the decision, ruling that although there was a constitutional right to refuse life-sustaining treatment, the states could regulate evidentiary requirements when there was no advance care directive.[15] Cruzan's family was eventually able to terminate her life support by moving her to a different hospital that would allow it. The *Cruzan* ruling made it appear that making one's treatment preferences very clear, ideally in written form, is the best protection against the chance of being condemned to a prolonged and needlessly technological or undignified end.

An advance directive is a formal document designed to provide such protection. Immediately after the highly publicized *Cruzan* case, interest in advance directives increased. In response to incidents like the *Cruzan* case, Congress passed the 1991 Patient Self-Determination Act (PSDA), which requires all health care agencies that receive Medicare or Medicaid funds to ask patients if they have advance directives or if they want educational materials on the subject. However, available evidence indicates that the PSDA did not increase the number of patients using advance directives.[16]

The fact that the number of Americans with advance directives hovers around only 20 percent limits the total impact of advance directives. There is also evidence that doctors fail to pay sufficient attention to documented patient preferences. SUPPORT, a study of end of life care at five geographically diverse

[15] *Cruzan by Cruzan v. Director.*

[16] E.J. Emanuel et al., "How well is the Patient Self-Determination Act working?"; Teno et al.

hospitals, showed that fewer than half of the physicians participating knew when their patients preferred not to be resuscitated.[17]

Even though advance directives have had a limited total impact on end-of-life care, they can still be valuable protection for the many individuals who are concerned about being subjected to invasive final care. Worries about the effectiveness of directives—particularly when circumstances come up that are not specifically addressed by the directive—are reasonable but can be addressed through the model of advance care planning. If the advance directive is only one component in a thorough process of communication over time between a patient, the doctor, and the potential proxy decision makers, then it is much more likely that the doctor or proxy will have the knowledge and the confidence to carry out the patient's wishes *in extremis*.[18]

Hospice

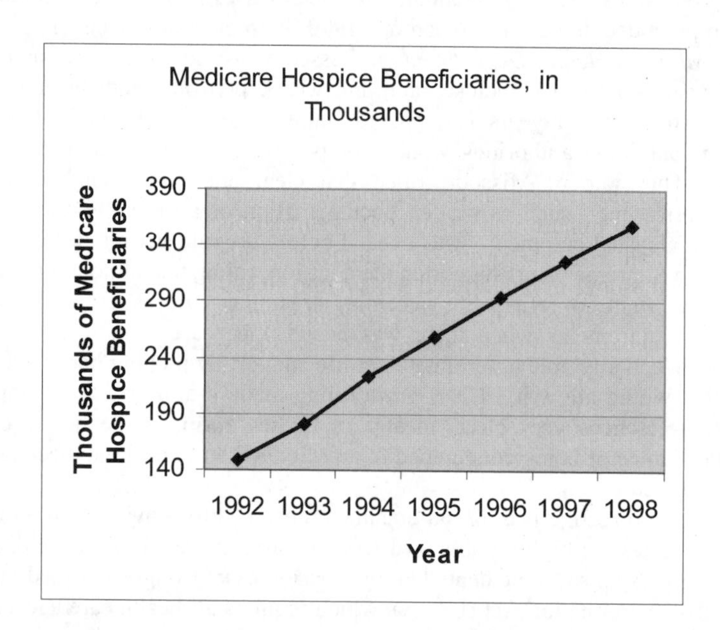

Figure 4. Number of People Using the Medicare Hospice Benefit Each Year. Based on data presented in the General Accounting Office Testimony Before the Special Committee on Aging, United States Senate. "Medicare: More Beneficiaries Use Hospice; Many Factors Contribute to Shorter Period of Use." Accessed at http://www.gao.gov/new.items/he00201t.pdf.

Hospices are designed to provide professional palliative care at the patient's own home or occasionally at a specialized facility. The first hospice in America was

[17] SUPPORT Principal Investigators.

[18] L.L. Emanuel et al., "Advance Care Planning."

founded in 1974, and the program received its major impetus in 1983 with the creation of the Medicare hospice benefit.[19] Under the benefit, if a Medicare patient's doctor believes the patient will not live more than six months, then Medicare will fund enrollment in a hospice. In just three decades, hospice has become so widely accepted that today roughly a fifth of all Americans die while enrolled in a hospice program.[20] The growth of hospice in the United States has been steady and rapid (Figure 4).

Five percent of a hospice's patient care hours must be provided by volunteers, and 80% of its patients must be outpatients. All these strictures are meant to encourage some critical aspects of a "good death."[21] A more controversial stricture is that if a patient opts to enroll in the hospice program, regular Medicare benefits are suspended, so that purely curative regimens are no longer covered. While a decision to enroll in hospice can be reversed, the enrollment is still a major decision that puts much treatment temporarily out of reach, and implies acceptance that death cannot be much delayed.

The original mission of hospice was primarily to care for cancer patients. Although hospice now serves patients with a variety of illnesses, cancer and hospice are still intimately linked. Data on Medicare recipients in Massachusetts and California showed that 33% of cancer decedents in Massachusetts and nearly 50% of cancer decedents in California enrolled in hospice. In both states, the majority of hospice enrollees had cancer—70% in Massachusetts and 60% in California.[22]

The place of death profile for hospice enrollees is totally different than for the average population. Hospice enrollees are three to five times more likely than non-enrollees to die at home (Table 2). So, while hospice is probably not the whole explanation for the change in place of death, it is plausible that it has significantly influenced the trend to out-of-hospital deaths. For instance, it is probably not a coincidence that cancer deaths have had by far the most dramatic shift out of hospital. It is also worth noting that use of hospice has grown rapidly throughout the period in which hospital deaths have declined. With current annual enrollment of over four hundred thousand people, hospice is large enough to account for a substantial part of the decrease in hospital death.

[19] Field and Cassel, eds.
[20] Christakis, "Predicting Patient Survival."
[21] Hoyer.
[22] E.J. Emanuel et al., "Managed Care, Hospice Use, Site of Death and Medical Expenditures."

Table 2. Percentage of Medicare Enrollees Dying in Each Place of Death (for hospice versus non-hospice enrolees in Massachusetts and California). Reprinted from E.J. Emanuel et al., "Managed Care, Hospice Use, Site of Death, and Medical Expenditures in the Last Year of Life." *Archives of Internal Medicine* 162 (15) 2002: 1722-8. Copyrighted © 2002, American Medical Association. All Rights Reserved.

	Massachusetts		California	
	Hospice	*Non-Hospice*	*Hospice*	*Non-Hospice*
Hospital Inpatient	10.6%	43.4%	5.2%	42.7%
Home	61.9%	12.7%	57.9%	17.5%
Nursing Home	24.2%	35.8%	29.3%	29.4%

While hospice is generally accepted as a good thing for patients, the statistical evidence that it improves quality of care is quite limited. What data there are indicate that hospice is effective in satisfying the desire of Americans to die at home, and that it may offer better pain management than conventional care does.[23] However, of the two studies supporting this claim, one is over fifteen years old and shows no dramatic difference, and the other is limited to patients in selected nursing homes. There is not currently comprehensive national data on the quality of hospice care and how it compares to care of the dying in hospitals.

Nursing Homes and Non-Hospice Home Care

A fifth of the whole population now dies in a nursing home, and the proportion has been rising steadily—in the 1990s, it increased from 16% to 22%.[24] Part of the explanation for this rise may be that nursing homes have become more likely to care for dying residents rather than transferring them to a hospital at the very end. Another striking trend associated with nursing homes is that people become more likely to die in a nursing home as they get older. The odds of death in a nursing home roughly double with each passing decade after 55, until 43% of people older than 85 die in one (Figure 5). Together, these trends mean that nursing homes are responsible for end-of-life care for a substantial segment of the overall population, and are of even more central importance to the dying of the very old.

[23] Wallston et al.; Miller et al.

[24] Flory et al.

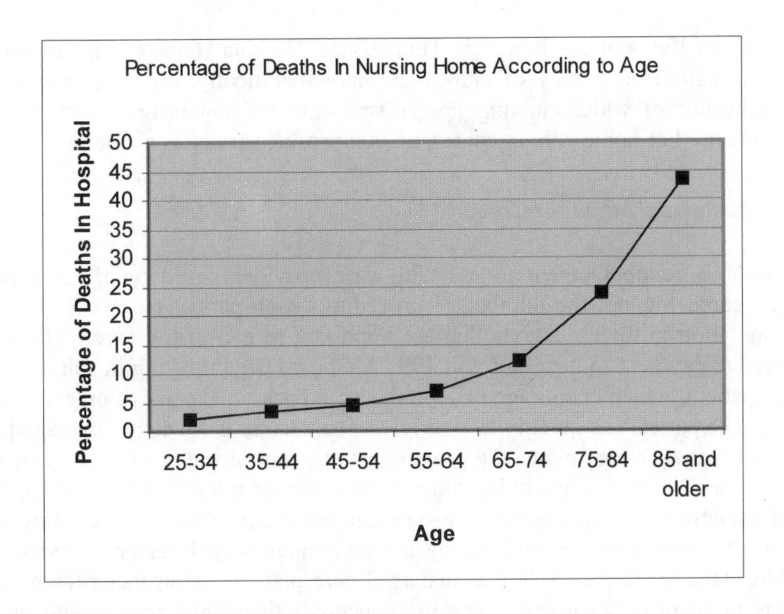

Figure 5. Percentage of Deaths Occuring in a Nursing Home, by age. Based on analysis of National Vital Statistics System death certificate records.

Home-care services are another aspect of elder care relevant to end of life. Medicare covers home care for beneficiaries who are housebound and need skilled nursing care, physical therapy, or speech therapy. The number of home health care patients increased very rapidly in the early 1990s, growing from 1.2 million in 1992 to 1.8 million in 1998, although the service has actually declined from a peak of 2.4 million patients in 1996.[25] Since they provide care for such large numbers of sick, largely elderly patients, it is almost certain that home-care services play a significant role in providing care at the end of many people's lives. They may have been a key factor in enabling the modern increase in home deaths.

However, not all insurance companies will cover such care, and not all patients and families will be able to afford the out-of-pocket costs of home care. Indeed, significant amounts of home care are provided, unpaid, by family members. One study of the impact of terminally ill patients on their families found that over a third of such patients required a "large amount of caregiving assistance" from a family member; and for 20% of patients, a family member took time off from work or quit their job to provide such care.[26] It is difficult to assess the economic burden of this care, but one important point to keep in mind is that growth in services like paid home-care, hospice, and nursing homes may relieve this hidden economic burden.

Both nursing homes and home care agencies are increasingly important to end-of-life care, but the quality of care given by each is hard to assess. The Medicare website provides a potentially useful quality assessment of over 17,000 Medicare and

[25] Centers for Disease Control and Prevention, National Center for Health Statistics.

[26] Covinsky et al.

Medicaid-certified nursing homes.[27] This service, Nursing Home Compare, does not give information on quality of end-of-life care specifically, but it is potentially a good indicator of which nursing homes will and will not provide an acceptable environment. For home-care agencies, no comparable service is available.

Capitation and Managed Care

Capitation is reimbursement to a health care provider based on the number of patients cared for, not the number of procedures each patient received. Medicare uses capitation in three contexts that are important to end-of-life care. The first is prospective payment to hospitals: in 1983 Medicare began paying hospitals a fixed fee per patient, which encouraged hospitals to cut back on expensive procedures and to discharge patients as quickly as possible. The second is Medicare managed care: like other managed care plans, the program pays a centralized health care provider a capitated rate, instead of using the older fee-for-service model. The third important use of capitation is hospice: hospices are reimbursed at a flat rate according to the number of patients they enroll, not by the individual procedures and services they provide. One consequence is that managed care plans have an incentive to enroll patients in hospice, because hospice then becomes financially responsible for care and the managed care plan does not have to pay for the high medical expenses of the last year of life, while receiving a small administrative fee.

Like hospice, managed care and prospective payment both discourage subjecting patients to unnecessary procedures and encourage out of hospital care. Spending less time in the hospital and undergoing fewer procedures could result in less invasive end-of-life care and a better chance of dying at home. On the other hand, capitation could also create an incentive to hold back on expensive procedures with palliative effects that dying patients might actually benefit from, such as radiation therapy. Research has been inconclusive on what the effect of capitation on place of death and the overall quality of dying has actually been.

The lack of satisfying information is due to a general dearth of information on the quality of end-of-life care, and in particular to the absence of data that isolates the effect of prospective payment and managed care. Prospective payment was introduced simultaneously in all but four states (Massachusetts, New Jersey, New York, and Maryland). A study of early 1980s trends in place of death at the time of prospective payment's introduction suggests that capitation may have acted to shift the place of death from hospital to nursing home, but there is insufficient information to make this claim definitively.[28] It is tempting to think that much of the decline in hospital death since the early 1980s is related to prospective payment, but the hypothesis has not been tested.

Managed care has been more extensively studied, and studies suggest that Medicare managed care increases hospice use by as much as 50%.[29] It is also

[27] Department of Health and Human Services.

[28] Sager et al.

[29] E.J. Emanuel et al., "Managed Care, Hospice Use, Site of Death, and Medical Expenditures."

associated with a higher proportion of home deaths and a lower proportion of nursing home deaths. Yet the same data show that Medicare managed care's impact on the rate of in-hospital deaths was very slight. The real question is whether managed care provides good end-of-life care, and the limited available data indicate that it is roughly as good as fee-for-service.[30] Convincing data on the effect of managed care on quality of dying still do not exist.

VARIATIONS IN END OF LIFE

The various changes to the medical system described above frequently affect various segments of the population differently, and the general trend away from dying in the hospital also varies among population segments. Age, race, and even geographic location affect the kinds of services a dying patient is likely to receive. A 60-year-old African-American male dying of cancer in the Eastern region of the United States is far more likely to die in the hospital than an 85-year-old white male dying of the same condition in the Western region.

Age

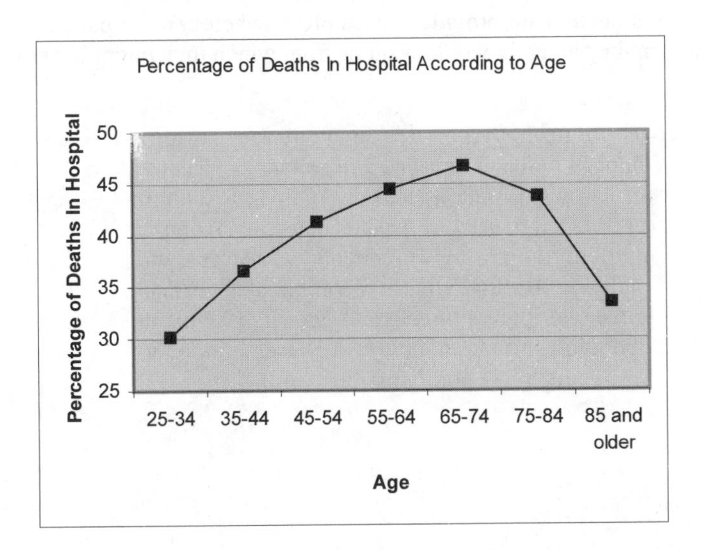

Figure 6. Percentage of Inpatient Deaths by Age Category. Based on data from J. Flory et al., "Place of Death: U.S. Trends Since 1980," *Health Affairs* 23 (2) 2004: 194-200. Reprinted with permission of the publisher: *Health Affairs*, http://www.healthaffairs.com.

[30] Slutsman et al.

Place of death is affected by the age of the decedent. In 1998 the proportion of in-hospital deaths peaked at about 47% for people aged 65-74. The proportion of in-hospital deaths then declined with age, down to 33% for people older than 85 (Figure 6). The lower rate of hospital mortality associated with greater age suggests that medicine takes a less interventionist approach to the care of the very old at the end of life.

This hypothesis is supported by a study of Medicare recipients in California and Massachusetts that indicates that the amount of money spent in the last year of life declined with age, as did the frequency of hospital admissions and aggressive medical interventions.[31] These age trends persisted for all major causes of death, even when controlling for comorbidity, which means that the difference in the way older people are treated is not explained by differences in cause of death between them and younger people. It appears that even when the likelihood of survival is the same, doctors are inclined to treat older people less aggressively. For instance, a survey of doctors' reactions to hypothetical cases showed that a third of the respondents would intubate a 40 year old male, if this gave him a 5% chance of survival, but they would not intubate an 80 year old male, even though the older man would have the same 5% chance of living.[32]

This bias is not necessarily inappropriate. The SUPPORT study, which was limited to severely ill patients in hospitals, found that older patients are less inclined to want aggressive therapy.[33] However, it may be that some age-based rationing by physicians and health care providers takes place, wherein older patients receive less treatment than they actually would want or from which they might benefit.

[31] Levinsky et al.
[32] Johnson and Kramer.
[33] Hamel et al.

Race and Ethnicity

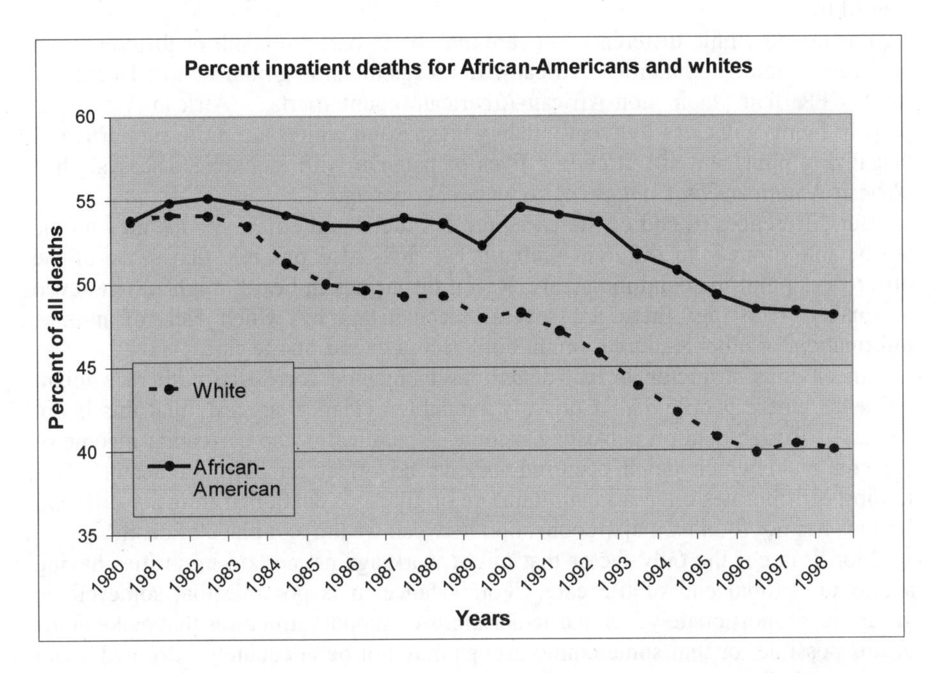

Figure 7. Percentage of African-Americans and Whites who Died as Hospital Inpatients. Reprinted from J. Flory et al. "Place of Death: U.S. Trends Since 1980," *Health Affairs* 23 (2) 2004: 194-200. Reprinted with permission of the publisher: *Health Affairs,* http://www.healthaffairs.com.

Place of death patterns for African-Americans and whites have diverged strikingly. In 1980, African-Americans and whites died in hospital at the same rate, but by 1998 whites were dying in the hospital significantly less often than African-Americans (Figure 7). More limited data suggest that other ethnicities—Asians and Hispanics—may also have a higher likelihood of dying in the hospital than whites or even African-Americans.[34] These differences in place of death persist even when cause of death is controlled for.

Differences in the end of life care for different ethnic groups go beyond place of death statistics. For instance, non-white racial and ethnic groups—African-Americans, Asians, and Hispanics—are underrepresented as hospice enrollees.[35] Another example is that, among Medicare recipients in Massachusetts and California, the medical expenses in the last year of life appear to be much higher—on the order of $10,000 more—for African-Americans than for whites. The higher

[34] Hempstead.
[35] National Hospice and Palliative Care Organization.

expense goes along with a higher use of expensive hospital procedures in the last year of life.[36]

Racial and ethnic differences in care may be in part the result of differences in preferences among ethnicities. African-Americans in nursing homes more frequently want CPR than their non-African-American counterparts;[37] African-Americans prepare living wills less frequently than whites,[38] and whites are more supportive of "legalizing physician aid in ending lives of patients with incurable diseases" than African-Americans by a margin of about twenty percent.[39]

But differences in end of life care between racial and ethnic groups may not be attributable entirely to different preferences. It is also possible that some of the differences point to social inequality. Racial disparities in health might extend even to quality of dying: there are several mechanisms by which lack of money, information, or other resources might limit access to end-of-life care.

An obvious suspicion is that disparities in income correlated with race might influence where people die. Currently available evidence suggests that this is not the case. Rough data on per-capita income—specifically, the per-capita income of the county in which death occurred—shows no effect on place of death. This finding corroborates international literature on place of death that in many different settings has not found a major connection between wealth and place of death.[40]

Money is not the only factor that might work against certain minorities having access to optimal end of life care. For instance, it is possible that some ethnic groups disproportionately lack particular kinds of support structures that make home deaths possible, or that some ethnic groups may not be adequately informed about hospice and similar programs. Palliative care programs do not have a representative share of physicians who are themselves minorities.[41] Since minority physicians disproportionately serve minority patients, this means that palliative care options may not be adequately available to some racial groups.

Because the factors that limit access to the best end of life care can be subtle, it is hard to separate preferences from other determinants. If a dying patient from an ethnic group in which hospice is an unfamiliar concept does not pursue the possibility of enrolling in hospice, is that because she really does not want that sort of care, or because she has not had a fair opportunity to learn about it?

[36] E.J. Emanuel et al., "Managed Care, Hospice Use, Site of Death, and Medical Expenditures"; Hogan et al.

[37] O'Brien et al.

[38] McKinley et al.

[39] Harvard School of Public Health/*Boston Globe*.

[40] Grande et al.

[41] Crawley et al.

Geography

There are also geographic differences in the distribution of place of death. The differences can be extreme: the rate of deaths in acute care hospitals in Washington D.C. was 73% in 1997; in Oregon, 32%.[42] The trend is for the proportion of in-hospital deaths to be highest in the Northeast and South, and lowest in the West and Midwest.[43]

One influential hypothesis has been that the kind of health care that patients receive is influenced, sometimes determined, by the kind of facilities that are locally available. The difference will apply even if the patients' needs and abilities to pay are the same. For example, one can imagine that people in a particular county that has several hospice facilities might be relatively likely to die in hospice, while an equivalent patient in another county could end up dying in the hospital just because there were no hospice facilities within convenient distance of her home.

One study indicates that the number of hospital beds in a given county has a significant influence on place of death; the more beds per thousand people, the greater the proportion of deaths occurring in-hospital.[44] This would suggest the plausible hypothesis that a factor as simple as the availability of hospital beds influences treatment decisions and discharge policy in ways relevant to end-of-life care. Another possibility is that number of hospital beds tracks a different variable that is actually responsible for the effect on place of death. Further research on this promising line of inquiry should show that some geographic variables—perhaps including local number of hospital beds and number of hospice facilities—significantly affect end-of-life care. It may prove important to make sure that facilities like hospice and hospitals are more equally distributed geographically.

Resource availability may not completely explain the differences in end-of-life care among regions. It could also be that the style of medical care in different parts of the country is actually different. Study of all these factors, from tangible differences like the number of hospital beds per thousand people to less easily measured variations in the style of care, may give policymakers an opportunity to pick out and promote the best aspects of different systems.

CONSIDERATIONS FOR IMPROVEMENT

Too Much Speculation, Too Little Data

Much good research has been done on end-of-life care, but much of it also points to the need for more data. Throughout this article we have used place of death as an indicator of changes in end-of-life care, but place of death alone cannot show whether the change was good or bad for terminally ill patients. To identify and fix problems in end-of-life care, information on quality of dying is indispensable. It is

[42] Hansen, Tolle, and Martin.
[43] Based on analysis of National Vital Statistics System death certificate records
[44] Wennberg and Cooper.

also in perpetually short supply, especially in three areas: nursing homes, hospice, and home care.

One of the most useful projects would be a study of the quality of dying in nursing homes. Researchers have made some progress in developing qualitative, sometimes quantitative, understanding of what death in a hospital is like. With so many people dying in nursing homes, we need similar information about them. Studies could highlight the strengths and deficiencies of end-of-life care in the nursing home, suggesting priorities for improvement. They would also be a step towards providing national quality data on end-of-life care in the nursing home. Nursing Home Compare, the web-based Medicare service, is a good model for how information on quality can be gathered and distributed, but it does not directly address the quality of care for the dying in a given nursing home.

Similarly, the quality of death in hospices across the U.S. is not well monitored, so work on a national scale to find out how well hospice enrollees are cared for compared to other dying patients would be very useful. Today, there is a good deal of confidence in the advantages of hospice, but that confidence is not currently backed up by substantial, reliable, and valid data. This is especially true because the sheer number of hospice patients is growing so quickly, and we should not take for granted that quality of care is being maintained. Data on quality should be gathered systematically, making it possible to rank hospices for purposes of quality improvement and patient choice. The quality of care available in hospices and nursing homes is rapidly coming to have a great impact on the well-being of many dying people. It is crucial to have mechanisms to monitor the quality of these services.

We know even less about the quality of at-home care than we do about hospice and nursing homes. Again, further study would be of dual use: first, as a guideline for improvements to care, and second, as a source of information on quality for families that use at-home care. At-home care does not simply include paid services, but also includes significant unpaid care given by family members.

Finally, studies of quality of dying that cover each geographic region of the country would be helpful in assigning meaning to the geographic variation in place of death. Different parts of the country have different ways of caring for the dying, and data on quality might help to identify any regions that have adopted particularly effective policies.

Equity

The variation in where people of different ages and different races are cared for at end of life is another area where more research would be useful. A major goal of such study should be to determine whether variation is due to legitimate causes, such as varying preferences, or troubling causes, such as poor access to home caregivers, hospice, or other factors that can provide better care.

Access of the kinds described above is something the health care system should strive to provide to all dying people equally. There is, at the least, no excuse for certain groups to be ignorant of their options and entitlements. For other factors, including the availability of home caregivers, the health care system's responsibilities are more ambiguous. For people without thorough health insurance or good volunteer care from family or community, home death can be very difficult and expensive, and nursing home or in-hospital death may become the only real options unless some agency will pay for home care. But making such caregivers available is not clearly in the purview of insurers or government. To what extent should society step in to help those who lack the social resources to die at home? It is possible that for many candidates for extensive home care, the nursing home is a more cost-effective but less appealing option: would it be permissible for insurance systems to save money by encouraging nursing home death instead?

Strained Budgets

The focus and hope of work on end-of-life care should be to improve care, not to improve medicine's financial bottom line. Yet it has long been hoped that the move away from the hospital would not just benefit patients, but also save enough billions of dollars to curb the increase in health care spending.

However, there is no evidence that the trend towards home death has saved or will save any significant sum of money.[45] End-of-life care has accounted for roughly the same overall percentage of Medicare spending—about 27%—since the 1970s.[46] The overall experience has been that improving care is possible, and limited savings are available sometimes, but massive savings have never materialized. For example, hospice was created with the hope that it would spend less per patient than regular Medicare did, but in the long run it has not done so by any significant margin. Part of the explanation for the lack of savings is the short time for which patients use hospice. In 1998, the median length of enrollment was thirty-six days.[47] Hospice may have the potential to be much cheaper than regular care, at least in the very last few months of life. But most enrollees use hospice so briefly that the savings from hospice have little time to add up to anything substantial.

Increased use of hospice—especially in the form of longer stays—might eventually save Medicare some money. However, the savings are unlikely to be very large, and they are by no means guaranteed. Hence policy makers should not depend on end-of-life care to resolve financial problems in the health system. On the other hand, it does appear that innovation in end-of-life care is not necessarily any more expensive than regular care: it is possible to make dying better without making it more expensive.

[45] Lubitz and Riley; E.J. Emanuel and L.L. Emanuel, "The illusion of cost savings."

[46] Lubitz and Riley.

[47] Christakis and Escarce, "Survival of Medicare Patients." Current estimates suggest even lower average length of stay.

CONCLUSION

As the baby boom generation passes the age of sixty-five, concern over both the quality and cost of end-of-life care is likely to increase. The experience of the last thirty years indicates that end-of-life care is an evolving rather than static field. The cost savings some hoped for have not materialized, but in other ways care for the dying in America have changed quite substantially, with hospitals playing less of a role and services like nursing homes and at-home care becoming more and more important. These changes affect hundreds of thousands of decedents and their families every year. Making services of adequate quality available to as many of these people as possible is a critical task.

REFERENCES

Bertman, Sandra. *Ars Moriendi: Illuminations on 'The Good Death' From the Arts and Humanities.* In *A Good Dying: Shaping Health Care for the Last Months of Life,* edited by J.K. Harrold and J. Lynn. New York: The Haworth Press, 1998.

________. *Facing Death: Images, Insights and Interventions.* New York: Routledge/Taylor and Francis Books, 1991.

Cantor, L.N. *Legal Frontiers of Death and Dying.* Bloomington: Indiana University Press, 1987.

Centers for Disease Control and Prevention, National Center for Health Statistics. *Trends from 1992, 1994, 1996, 1998, and 2000.* Available at: http://www.cdc.gov/nchs/about/major/nhhcsd/nhhcsd.htm (accessed July 6, 2004).

Christakis, N.A. "Predicting Patient Survival Before and After Hospice Enrollment." In *A Good Dying: Shaping Health Care for the Last Months of Life,* edited by J.K. Harrold and J. Lynn. New York: The Haworth Press, 1998.

Christakis, N.A. and J.J. Escarce. "Survival of Medicare Patients After Enrollment in Hospice Programs." *New England Journal of Medicine* 335, no. 3 (1996): 172-178.

Covinsky, K.E., L. Goldman, E.F. Cook, R. Oye, N. Desbiens, D. Reding, W. Fulkerson, A.F. Connors, J. Lynn, and R.S. Philips, for the SUPPORT Investigators. "The Impact of Serious Illness on Patients' Families." *Journal of the American Medical Association* 272 (1994): 1839-1844.

Crawley, L., R. Payne, J. Bolden, T. Payne, P. Washington, and S. Williams. "Palliative and End-of-Life Care in the African American Community." *Journal of the American Medical Association* 284 (2000): 2518-21.

Cruzan v. Director. Vol. 497 U.S. 261 (1990).

Department of Health and Human Services. "Nursing Home Overview." Available at: www.medicare.gov/nursing/overview.asp (accessed July 6, 2004).

Emanuel, E.J. "Why Now?" In *Regulating How We Die*, edited by E. Emanuel. Cambridge: Harvard University Press, 1998.

Emanuel, E.J., A. Ash, W. Yu, G. Gazelle, N.G. Levinsky, O. Saynina, M. McClellan, and M. Moskowitz. "Managed Care, Hospice Use, Site of Death, and Medical Expenditures in the Last Year of Life." *Archives of Internal Medicine* 162, no. 15 (2002): 1722-8.

Emanuel, E.J. and L.L. Emanuel. "The illusion of cost savings at the end of life." *New England Journal of Medicine* 330 (1994): 540-544.

Emanuel, E.J., D.S. Weinberg, R. Gonin, L.R. Hummel, L.L. Emanuel. "How well is the Patient Self-Determination Act working? an early assessment." *American Journal of Medicine* 95 (1993): 619-28.

Emanuel, L.L., M. Danis, R.A. Pearlman, and P.A. Singer. "Advance Care Planning as a Process: Structuring the Discussions in Practice." *Journal of the American Geriatrics Society* 43 (1995): 440-446.

Field, M.J. and C.K. Cassel, eds. *Approaching Death: Improving Care at the End of Life*. Washington, DC: National Academy Press; 1997.

Flory J., Y. Young-Xu, I. Gurol, N. Levinsky, A. Ash, and E. Emanuel. "Place of Death: U.S. Trends Since 1980." *Health Affairs* 23, no. 2 (2004): 194-200.

Gallup Organization. *Knowledge and attitudes related to hospice care. Survey conducted for National Hospice Organization.* Princeton: The Gallup Organization, 1996.

Grande, G.E., J.M. Addington-Hall, and C.J. Todd. "Place Of Death And Access To Home Care Services: Are Certain Patient Groups At A Disadvantage?" *Social Science and Medicine* 47 (1998): 565-579.

Hamel, M.B., J. Lynn, J.M. Teno, K.E. Covinsky, A.W. Wu, A. Galanos, N.A. Desbiens, and R.S. Phillips. "Age-related differences in care preferences, treatment decisions, and clinical outcomes of seriously ill hospitalized adults: lessons from SUPPORT." *Journal of the American Geriatrics Society* 48 (2000): S176-S182.

Hansen, S.M., S.W. Tolle, and D.P. Martin. "Factors Associated with Lower Rates of In-Hospital Death." *Journal of Palliative Medicine* 5, no. 5 (2002): 677-85.

Harvard School of Public Health/*Boston Globe*. *National Attitudes Towards Death and Terminal Illness*. Needham: KRC Communications Research, 1991.

Hempstead, K. "Trends in Place of Death in New Jersey: An Analysis of Death Certificates." *Topics in Health Statistics*. Trenton: New Jersey Department of Health and Senior Services, 2001.

Hogan, C., J. Lunney, J. Gabel, J. Lynn. "Medicare beneficiaries' costs of care in the last year of life." *Health Affairs* 20 (2001): 188-95.

Hoyer, T. "A History of the Medicare Hospice Benefit." In *A Good Dying: Shaping Health Care for the Last Months of Life,* edited by J.K. Harrold and J. Lynn. New York: The Haworth Press, 1998.

In Re Quinlan Vol. 70 N.J. 10 (1976)

Johnson, M.F. and A.M. Kramer. "Physicians' responses to clinical scenarios involving life-threatening illness vary by patients' age." *Journal of Clinical Ethics* 11 (2000): 323-327.

Levinsky, N.G., W. Yu, A. Ash, M. Moskowitz, G. Gazelle, O. Saynina, and E. J. Emanuel. "Influence of Age on Medicare Expenditures and Medical Care in the Last Year of Life." *Journal of the American Medical Association* 286 (2001): 1349-1355.

Lubitz, J.D. and G.F. Riley. "Trends in Medicare payments in the last year of life." *New England Journal of Medicine* 328 (1993): 1092-1096.

McKinley, E.D., J.M. Garett, A.T. Evans, and M. Danis. "Differences in end-of-life decision making among African-American and white ambulatory cancer patients." *Journal of General Internal Medicine* 11 (1996): 651-656.

Miller, S.C., P. Gozalo, V. Mor, and C. Susan. "Outcomes and Utilization for Hospice and Non-Hospice Nursing Facility Decedents." Available at: http://aspe.hhs.gov/daltcp/reports/oututil.pdf (accessed July 6, 2004).

Minino, A.M., E. Arias, M.D. Kochanek, S.L. Murphy, and B.L. Smith. "Deaths: Final Data for 2000." *National Vital Statistics Reports* 50, no. 15 (2002).

National Hospice and Palliative Care Organization. *NHPCO's Facts and Figures on Hospice and Palliative Care*. Available at http://www.nhpco.org/i4a/pages/Index.cfm?pageid=3362 (accessed July 6, 2004).

O'Brien, L.A., J.A. Grisso, G. Maislin, K. LaPann, K.P. Krotki, P.J. Greco, P.E.A. Siegert, and L.K. Evans. "Nursing Home Residents' Preferences for Life-Sustaining Treatments." *Journal of the American Medical Association* 274 (1995): 1775-1779.

Pritchard R.S., E.S. Fisher, J.M. Teno, S.M. Sharp, D.J. Reding, W.A. Knaus, J.E. Wennberg, and J. Lynn. "Influence of Patient Preferences and Local Health System Characteristics on the Place of Death." *Journal of the American Geriatics Society* 46 (1998): 1242-1250.

Sager, M., D.V. Easterlin, D.A. Kindig, O.W. Anderson. "Changes in the Location of Death After Passage of Medicare's Prospective Payment System." *New England Journal of Medicine* 320 (1989): 433-439.

Slutsman, J., L.L. Emanuel, D. Fairclough, D. Bottorff, and E.J. Emanuel. "Managing end-of-life care: comparing the experiences of terminally ill patients in managed care and fee for service." *Journal of the American Geriatrics Society* 50 (2002): 2077-83.

SUPPORT Principal Investigators. "A Controlled Trial to Improve Care for Seriously Ill Hospitalized Patients. The Study to Understand Prognoses and Preferences for Outcomes and Risks of Treatments." *Journal of the American Medical Association* 274 (1995): 1591-1598.

Teno, J., J. Lynn, N. Wenger, R.S. Phillips, D.P. Murphy, A.F. Connors Jr., N. Desbiens, W. Fulkerson, P. Bellamy, and W.A. Knaus. "Advance Directives for Seriously Ill Hospitalized Patients: Effectiveness with the Patient Self-Determination Act and the SUPPORT Intervention." *Journal of the American Geriatrics Society* 45 (1997): 500-507.

Vacco, Attorney General of New York v. Quill. Vol. 117 S.Ct. 2293 (1997)

Wallston, K.A., C. Burger, R.A. Smith, and R.J. Baugher. "Comparing the Quality of Death for Hospice and Non-Hospice Cancer Patients." *Medical Care* 26, no. 2 (1988): 177-182.

Washington v. Glucksberg. Vol. 521 U.S. 702 (1997).

Wennberg, J. and M. Cooper. *The Dartmouth Atlas of Health Care in the United States.* Chicago: American Hospital Publishing, 1998.

III.

ENVIRONMENTAL ETHICS

J. BAIRD CALLICOTT

THE PRAGMATIC POWER AND PROMISE OF THEORETICAL ENVIRONMENTAL ETHICS

Forging a New Discourse[1]

Abstract. Pragmatist environmental philosophers have (erroneously) assumed that environmental ethics has made little impact on environmental policy because environmental ethics has been absorbed with arcane theoretical controversies, mostly centered on the question of intrinsic value in nature. Positions on this question generate the allegedly divisive categories of anthropocentrism/ nonanthropocentrism, shallow/ deep ecology, and individualism/ holism. The *locus classicus* for the objectivist concept of intrinsic value is traceable to Kant, and modifications of the Kantian form of ethical theory terminate in biocentrism. A subjectivist approach to the affirmation of intrinsic value in nature has also been explored. Because of the academic debate about intrinsic value in nature, the concept of intrinsic value in nature has begun to penetrate and reshape the discourse of environmental activists and environmental agency personnel. In environmental ethics, the concept of intrinsic value in nature functions similarly to the way the concept of human rights functions in social ethics. Human rights has had enormous pragmatic efficacy in social ethics and policy. The prospective endorsement of the Earth Charter by the General Assembly of the United Nations may have an impact on governmental environmental policy and performance similar to the impact on governmental social policy and behavior of the adoption by the same body in 1948 of the Universal Declaration of Human Rights. Belatedly, but at last, the most strident Pragmatist critics of the concept of intrinsic value in nature now acknowledge its pragmatic power and promise.

INTRODUCTION

In one of the most ancient and venerable sources of Chinese philosophy, the *Analects,* a disciple asks Confucius what he would do first were he to become the prime minister of the State of Wei.[2] Without question, Confucius replies, first I would rectify names. His disciple was puzzled by this saying; and for a long time so was I. But no more, for I am coming to appreciate the power of names, and of discourse, more generally, in the formation of environmental policy.

The true answer to Juliet's question, "What's in a name?" in Shakespeare's play is "Really, quite a lot." Consider various names for women—"chicks," "babes," "broads," "ladies." The feminist movement has made us keenly aware that what we call someone or something—what we name her, him, or it—is important. A name frames, colors, and makes someone or something available for certain kinds of uses… or abuses. Even the name "lady" is freighted with so much baggage that it is not worn comfortably by many women. A major effort of feminist politics has been

[1] This essay was first published in *Environmental Values* 11 (2002): 3-25. Reprinted by permission of The White Horse Press, Isle of Harris, U.K.

[2] Hall and Ames.

A. W. Galston and C. Z. Peppard (eds.), Expanding Horizons in Bioethics, 185-208.

the rectification of names for women, and more generally, the rectification of gender discourse.

Self-styled Pragmatist environmental philosophers have complained that environmental philosophy has been bogged down in ivory-tower theorizing to little practical effect.[3] Here I argue that theoretical environmental philosophy has had and is having a profound, albeit indirect, practical effect on environmental policy. It has done so by creating a new discourse that environmental activists and environmental professionals have adopted and put to good use. At the heart of this new discourse is the concept of intrinsic value in nature. I sketch the history of this concept and its associated discourse, and indicate how it is practically impacting environmental policy.

ENVIRONMENTAL PHILOSOPHY: MORE THEORETICAL THAN APPLIED

Environmental philosophy has been less an "applied" subdiscipline of philosophy than some of the other applied subdisciplines with which it is often lumped-biomedical ethics, business ethics, and engineering ethics, for example. Environmental philosophy has, more particularly, been more involved with reconstructing ethical theory than with applying standard, off-the-rack ethical theories to real-world environmental problems.

In large part that is because standard ethical theory had been so resolutely—even militantly—anthropocentric that it seemed inadequate to deal with today's environmental problems. In scope and magnitude, contemporary human transformation of the environment is unprecedented. Gradually, the impact of human activities on nonhuman nature became almost ubiquitous in scope and unrelenting in intensity, so much so that by the mid-twentieth century, the existence of an environmental *crisis* was widely acknowledged. And the contemporary environmental crisis seems morally charged. For example, the current orgy of human-caused species extinction seems wrong—morally wrong. And not just because the anthropogenic extinction of many species might adversely affect human interests or human rights. Most first-generation environmental philosophers, therefore, took the task of environmental ethics to be constructing a nonanthropocentric theory of ethics that would somehow morally disenfranchise nonhuman natural entities and nature as a whole—directly, not merely indirectly to the extent that what human beings do in and to nature would affect human interests and human rights.

This was the burden of the first academic paper in the field, "Is There a Need for a New, an Environmental Ethic?" by Australian philosopher Richard Routley, presented to the Fifteenth World Congress of Philosophy in Varna, Bulgaria in 1973.[4] A similar task was set by Norwegian philosopher Arne Naess in his paper,

[3] Norton, "Epistemology and Environmental Values."
[4] Sylvan.

"The Shallow and the Deep, Long-Range Ecology Movements: A Summary."[5] In the first paper on environmental ethics by an American philosopher, Holmes Rolston III argued that the central task of environmental philosophy is to develop a "primary," not a "secondary," "ecological ethic."[6] Animal rights theorist Tom Regan reiterated Rolston's understanding of the enterprise—that a proper environmental ethic was "an ethic *of* the environment," not an "ethic for the *use* of the environment," which he called a mere "management ethic."[7]

THE KANTIAN CONCEPT OF INTRINSIC VALUE

Central to the theoretical challenge of developing a direct, primary ethic *of* the environment is the problem of intrinsic value in nature. Although the early twentieth-century English philosopher G.E. Moore wrote much about intrinsic value,[8] Immanuel Kant's modern classical concept of intrinsic value and the way it functioned in his ethics most influenced the thinking of contemporary environmental philosophers.[9] Central to Kant's ethic is the precept that each person be treated as an end in him- or herself, not merely as a means. Indeed, the second formulation of Kant's categorical imperative is this: "Act so that you treat humanity, whether in your own person or that of another, always as an end and never as a means only."[10] Kant justifies—or "grounds"—this precept by claiming that each person has intrinsic value. That claim in turn is justified by finding in each person an intrinsic value-conferring property, which Kant identified as reason. Thus, rational beings, according to Kant, have intrinsic value, and should therefore be treated as ends in themselves and never as means only.

This Kantian approach to ethics appears at first glance to be unpromising for developing a *nonanthropocentric* environmental ethic, as Routley, Naess, Rolston, and Regan so unambiguously set forth the task. Why? Because Kant's intrinsic value-conferring property, reason or rationality, had long been regarded as a hallmark of human nature. At the dawn of Western philosophy, Aristotle declared that reason or rationality was the "differentia" that distinguished "man," as a species, from the other animals. *Anthropos* is the uniquely "rational animal," according to Aristotle. Thus, Kant's approach to ethics appears to be a brief for anthropocentrism and to foreclose the possibility of nonanthropocentrism. Indeed, Kant goes out of his way to exclude non-human natural entities and nature as a whole from ethical enfranchisement:

> Beings whose existence does not depend on our will but on nature, if they are not rational beings, have only relative worth as means and are therefore called 'things'; on

[5] Naess.
[6] Rolston, "Is There an Ecological Ethic?"
[7] Regan, "The Nature and Possibility of an Environmental Ethic."
[8] Moore.
[9] Kant.
[10] Ibid., 39.

> the other hand, rational beings are designated 'persons' because their nature indicates that they are ends in themselves, i.e., things which may not be used as a means.[11]

For Kant, human beings are ends; beings whose existence depends on nature are means.

EXTENDING THE KANTIAN CONCEPT OF INTRINSIC VALUE TO (SOME) ANIMALS

But look again. In the *Foundations of the Metaphysics of Morals,* Kant himself is quite careful to avoid speciesism—analogous to racism and sexism—the *unjustified* or *ungrounded* moral entitlement of one's own kind and the exclusion of other kinds. Not being human, but being *rational* is that in virtue of which a human being has intrinsic value. Kant consistently holds open the possibility that there may be other-than-human rational beings. He never more specifically identifies who such nonhuman rational beings may be. Some passages suggest Kant might be thinking of God and the heavenly host; others that he might be thinking of rational beings on other planets that inhabit very different bodies and therefore have very different desires and inclinations than do human beings. In the passage just quoted, he seems to hold open the possibility that there may be nonhuman rational beings found in terrestrial nature. It is in this orthodox Kantian moral climate that so much ethical significance was recently attached to proving that chimpanzees and gorillas could master rudimentary language skills and could, via American Sign Language or some other surrogate for spoken language, express themselves creatively.[12] For Descartes had insisted that the ability to use language creatively—not merely rotely as he believed parrots to do—was an indication of rationality.[13]

Proving that chimpanzees and gorillas are minimally rational does undermine anthropocentrism, but only a little. It certainly does not take us very far in the direction of an expansive environmental ethic—however much it may help ethically rehabilitate our primate relatives and spare them the indignities and outrages of the zoo trade and biomedical research. Kant's conceptual distinction between humanity and rationality was, however, also exploited theoretically another way, which proved to be more powerful and transformative. Not all human beings are minimally rational. The so-called "marginal cases" are not.[14] Infants, the severely mentally handicapped, and the abjectly senile are the usual suspects. They are thus in the same boat with all the other "[b]eings whose existence… depend[s] on nature… i.e., things which may be used merely as a means," to quote Kant once more.[15] Let's get specific. If we equitably applied Kant's ethical theory, we could justifiably perform the same painful and destructive biomedical experiments on unwanted non-rational

[11] Ibid., 46.
[12] Savage-Rumbaugh, Shanker, and Taylor.
[13] Descartes.
[14] Regan, "Examination and Defense of One Argument Concerning Animal Rights."
[15] Kant, 46.

infants that we inflict on non-rational nonhuman animals; we could open up a hunting season on the severely mentally handicapped; and we could make pet food out of the abjectly senile.

Such abhorrent implications of Kant's moral philosophy provided nonanthropocentric theorists with an opportunity to propose retaining Kant's form of moral argument—which has, after all, been so compelling in Western ethical thought—but revising its specific conceptual contents, so as to include the marginal cases in the class of persons and rescue them from the class of things. The form or ethical architecture that was retained is Kant's close linkage of moral ends, intrinsic value, and a value-conferring property. Thus to be a moral end, and not a means only, you must have intrinsic value, but making rationality the value-conferring property, appears, in light of the "Argument from Marginal Cases," to be too restrictive. Various alternatives to rationality have been proposed, selected to justify the theorist's personal agenda. Regan, who was content to limit "moral considerability" to warm, furry animals, proposed being the "subject of a life" as the intrinsic value-conferring property.[16] Subjects of a life have a sense of self, remember a personal past, entertain hopes and fears about the future—in sum, enjoy a subjective state of being, which can be better or worse from their own point of view. Peter Singer, who wanted to extend "moral considerability" a bit more generously, proposed sentience, the capacity to experience pleasure and pain, as the intrinsic value-conferring property.[17] That move reached a much wider spectrum of animals—how wide is not completely clear—but, clearly, it left out the entire plant kingdom.

EXTENDING THE KANTIAN CONCEPT OF INTRINSIC VALUE TO ALL LIVING BEINGS

To reach out and touch all living beings with moral considerability, several theorists proposed having *interests* as a plausible and defensible intrinsic value-conferring property.[18] A living being—a tree for example—can have interests in the absence of consciousness. This basic idea was variously expressed. A living being has a good of its own, whether or not it is good for anything else. Unlike complexly functioning machines, such as automobiles, whose ends or functions are determined or assigned to them by their human designers to serve human ends, living beings have ends, goals, or purposes—*teloi,* in a word—of their own. They are, in Paul Taylor's terminology, "teleological centers of life."[19] In Warwick Fox's, they are *autopoietic:* self-creating and self-renewing.[20]

[16] Regan, *The Case for Animal Rights*.
[17] Singer, *Animal Liberation*.
[18] See, e.g., Goodpaster; Johnson; Taylor.
[19] Taylor.
[20] Fox.

PROBLEMS WITH BIOCENTRISM AND THEIR PROPOSED SOLUTIONS

The main problem, theoretically speaking, with biocentrism—as this modified or expanded Kantian approach to nonanthropocentric environmental ethics has come to be called—is that it seems to stop with individual organisms. Biocentrism distributes intrinsic value both too broadly *and* too narrowly.

As to the former, granting each and every organism moral considerability makes ethical space way too densely crowded, rendering our most routine and vital human actions ethically problematic. Surely, it is perfectly possible to refrain from ill-using our fellow primates as objects of amusement and subjects of medical experimentation, with little human inconvenience. Equally possible—and with only a little more mindfulness, abstemiousness, and inconvenience—we might give up eating meat and using other products made from animals, our fellow sentient beings. But we have to eat something, slap mosquitoes and other annoying insects, rid ourselves and our domiciles of vermin, weed our flower gardens—all of which are morally questionable if every living being has intrinsic value and should be treated as an end in itself, not a means only.

On the other hand, biocentrism too narrowly distributes intrinsic value in nature because it does not provide moral considerability for what environmentalists most care about. Frankly, environmentalists do not much care about the welfare of each and every shrub, bug and grub. We care, rather, about preserving *species* of organisms, *populations* within species, and *genes* within populations. In a word, we care about preserving biodiversity. We care about preserving communities of organisms and ecosystems. We also care about *air* and *water* quality, *soil* stability, and the integrity of Earth's stratospheric *ozone membrane.* None of these things appear to have interests, goods of their own, ends, purposes or goals, and thus none has intrinsic value, on this account.

Solutions to both biocentric distribution problems have been proposed. A solution to the too-broad distribution problem is to distribute intrinsic value unequally or differentially.[21] Grant all organisms baseline or minimal intrinsic value. Thus, when our own interests are not at stake, we should leave them alone to pursue their own ends, to realize their own *teloi,* each in its own way. Additional intrinsic value is distributed to sentient organisms, yet more to subject-of-a-life organisms, and more still to rational organisms.[22] Thus, because we human beings have the most intrinsic value, as rational and sentient subjects of a life, we are entitled to defend it and cater to it by doing bad things to other organisms with less intrinsic value—but only if we conscientiously deem it to be necessary. That seems plausible enough, although rather conventional, leaving us human beings at the top of the moral pyramid where we have always been. The difference is that in traditional Western ethics the pyramid was low and squat. Nonhuman organisms were mere things, with no intrinsic value at all. They were thus available for any human use at all, however fatuous. Differential biocentrism extends the moral

[21] Goodpaster.

[22] Rolston, *Environmental Ethics: Duties to and Values in the Natural World.*

pyramid's height and mass to much greater proportions, albeit leaving human beings at the pinnacle.

A solution to biocentrism's too-narrow distribution problem is less plausible. Lawrence Johnson has seized upon somewhat dated, minority views in evolutionary biology and ecology to argue that species and ecosystems have interests.[23] Some biologists have argued that species are not collections of organisms capable of interbreeding, but supra-individuals that are protracted in space and time.[24] If so, we may convince ourselves that they have interests, and therefore intrinsic value, and therefore moral considerability. And there is a long, albeit fading, tradition in ecology that conceives ecosystems to be superorganisms to which individual organisms are related as cells and species as organs.[25] And if so, again, we may believe that they have interests, and therefore intrinsic value, and therefore moral considerability. But these are big "ifs." Rolston takes a different approach. He points out that the most fundamental end of most organisms is to realize their genetic potential—to represent ("re-present") their species and reproduce ("re-produce") it.[26] They have a good of their own, *which is their species.* Thus does Rolston try to convince us that species per se may plausibly be said to have intrinsic value. For organisms to flourish, even to live at all, they must live in an ecological context or habitat. Thus does Rolston try to justify finding intrinsic value in biotic communities and ecosystems.

THE SUBJECTIVIST ACCOUNT OF INTRINSIC VALUE IN NATURE

This mainstream line of argument in environmental ethics, which begins with a Kantian superstructure, works through animal liberation, and terminates in biocentrism, assumes that intrinsic value supervenes or piggybacks on some objective property. Thus intrinsic value, albeit supervenient, itself therefore appears to be an objective property in nature. Indeed, the adjective "intrinsic" seems logically to require that *intrinsic* value, if it exists at all, exists as an objective property. It is intrinsic to the being that has it. Kant himself appears to think that intrinsic value is something objective: "Such beings [rational beings] are not merely subjective ends whose existence as a result of our action has a worth for us, but are objective ends, i.e., beings whose existence in itself is an end."[27] But the idea that value—or worth—of any kind can be objective seems to fly in the face of a shibboleth of modern Western philosophy: René Descartes' division of the world into the *res extensa* and the *res cogitans*, the subjective and objective domains, respectively, and David Hume's ancillary distinction between fact and value. *All* value is, from the most fundamental modern point of view, subjectively conferred.

[23] Johnson.
[24] See, e.g., Ghiselin; Hull.
[25] McIntosh.
[26] Rolston, *Environmental Ethics: Duties to and Values in the Natural World.*
[27] Kant, 46.

No valuing subject, no valuable objects. That is, without the existence of valuing subjects, no value of any kind would exist in the world—from a modern point of view.

Nevertheless, some nonanthropocentric environmental philosophers (I among them) have argued that a robust account of intrinsic value in nature can be provided even within the severe constraints of the allied object-subject/ fact-value distinctions.[28] From a modern point of view, "value" is first and foremost a verb. Value, more technically put, is conferred on an object by the intentional act of a valuing subject. If so, "instrumental" and "intrinsic" may be regarded as adverbs, not adjectives. Thus one may value (verb transitive) some things instrument*ally*—our houses, cars, computers, clothes and such. Similarly, one may value (verb transitive) other things intrinsic*ally*—ourselves, our spouses, children, and other relatives. If we have learned our religion and moral philosophy well, we may intrinsically value all other human beings. Indeed, it is logically possible to value intrinsically anything under the sun—an old worn out shoe, for example. But most of us value things intrinsically when we perceive them to be part of a community to which we also belong, because we are evolved to do so.[29]

"Perceive" is the key word here, for perception can be trained and redirected. Much of the suasive environmental literature aims to train and redirect our perception of nature such that we see it as the wider community in which all other communities are embedded. Aldo Leopold's *A Sand County Almanac* is an outstanding example. In the Foreword, Leopold writes, "We abuse land because we regard it as a commodity belonging to us. When we see land as a community to which we belong, we may begin to use it with love and respect."[30] Most of the remainder of the book is devoted to persuading us that ecology "enlarges the boundaries of the community to include soils, waters, plants, and animals, or collectively the land."[31] When that happens, people will have "love, respect, and admiration for land, and a high regard for its value… [and b]y value I mean something far broader than mere economic value; I mean value in the philosophical sense"—intrinsic value, in other words.[32]

THE CRYPTO-SUBJECTIVISM OF ALLEGEDLY OBJECTIVIST ACCOUNTS OF INTRINSIC VALUE

How could Kant, a thoroughly modern philosopher, and a close student of Hume, actually think that intrinsic *value* is an objective property (of rational beings)? A closer reading of Kant himself indicates that in fact he does not think it is. Kant writes, "Man necessarily *thinks* of his existence this way"—that is, as an end-in-

[28] See, e.g., Callicott, *Beyond the Land Ethic;* O'Neill.
[29] Callicott, *Beyond the Land Ethic.*
[30] Leopold, viii.
[31] Ibid., 204.
[32] Ibid., 223.

itself, something of intrinsic value—"Thus far, it is a *subjective* principle of human action."[33] Kant is intellectually honest; he is fully aware that, given the constraints of the Cartesian *res cogitans/res extensa* and ancillary Humean fact/ value distinctions, value is not objective in the same sense that a rock is objective, something existing independently of the intentional act of valuing a subject in the *res extensa.* Kant goes on, however:

> Also every other rational creature *thinks* of his own existence by means of the same rational ground which holds also for myself, thus it is at the same time an *objective* principle from which, as a supreme practical ground, it must be possible to derive all laws of the will.[34]

The meaning of "objective," in the above-quoted fragment from Kant, is "universal," not "existing independently of the intentional act of a valuing subject." In other words, Kant uses the concept of objectivity in its epistemological, not ontological, sense. Each organism should be an unconditional end for moral agents, because for itself it is an unconditional end-in-itself.

A closer reading of Rolston—the most subtle thinker of the purportedly objectivist school of intrinsic value-in-nature theorists—also shows that he follows Kant in effecting an unmarked shift from the ontological to the epistemic sense of "objective" and back again. We human beings self-consciously value ourselves, as well as other things, intrinsically. But lemurs, Rolston notes, also demonstrably value themselves intrinsically, although perhaps not self-consciously.[35] So do warblers. What Rolston does find in nature is a wide spectrum of nonhuman reflexively valuing subjects. He begins with human subjects, then moves on to our close relatives, phylogenetically speaking, and on from there, to subjects more distantly related and arguably less acutely conscious than lemurs and other primates—birds, reptiles, insects—all in some sense self-valuing subjects. Finally, Rolston posits the existence of valuing subjects stripped of all sensitivity: "Trees are also valuable in themselves," Rolston writes.[36] But why? How? Because, as he explains, they are "able to value themselves." In what sense? Is Rolston going beyond conventional science and claming a secret, inner life for plants? Not at all:

> Natural selection picks out whatever traits an organism has that are valuable to it, relative to its survival. When natural selection has been at work gathering these traits into an organism, *that organism is able to value* on the basis of those traits. *It is a valuing organism*, even if the organism is not a sentient valuer...[37]

So, clearly, although the valuing subject may lack sentience, indeed consciousness of any kind—that is, the valuing subject may, paradoxically, lack subjectivity—Rolston agrees with the subjectivists that the value of any object, a valu*ee*, depends, in the last analysis, on the existence of a valuing subject, a valu*er*.

[33] Kant, 47.
[34] Ibid., 47, emphasis added.
[35] Rolston, *Conserving Natural Value.*
[36] Rolston, "Naturalizing Callicott," 118.
[37] Ibid., 119.

For Rolston, the ethical payoff of this analysis is characteristically Kantian. Rolston's environmental ethics follows the Kantian pattern, but broadens the "subjective principle" to the maximum extent possible. Reflexive self-valuing is not confined to "man," nor to "rational creatures," nor even to sentient or conscious creatures, but to any and all evolved creatures. And, just as Kant, Rolston argues that because they value themselves intrinsically, we too should value them intrinsically. That makes the principle "objective," but in a different sense of the word, which neither Kant nor Rolston marks.

THE PRAGMATIST CRITIQUE OF THEORETICAL ENVIRONMENTAL PHILOSOPHY

As this brief summary will indicate—and, believe me, it is brief, sketchy, and incomplete, given the voluminous literature on the subject—mainstream environmental philosophy has been preoccupied with a very abstruse and arcane theoretical project. A growing cadre of environmental philosophers, identifying themselves as Pragmatists of one kind or another, has begun to protest against this preoccupation with theory, especially this theoretical problem of intrinsic value in nature.[38] They variously, but basically, argue that it makes no difference to environmental practice and policy whether we think of nature as having intrinsic value or only instrumental value. Whether we value nature as a means to human ends or an end in itself, we still value it—and therefore will save it. Norton calls this the "convergence hypothesis."[39] Because the concept of intrinsic value in nature makes no difference to environmental practice and policy, debate about it is a waste of time and intellectual capital that could better be spent on something more efficacious. Further, lay people cannot understand the jargon-ridden, abstract discourse of theoretical environmental philosophy. If they do get an inkling of what it is about, they will be alienated from it, because most lay people are uncritically anthropocentric. Worse, nonanthropocentrism and the concept of intrinsic value in nature is divisive, setting environmental philosophers at odds with one another, occasioning endless, unbecoming bickering between shallow and deep theorists, and among the deep theorists, between subjectivists and objectivists.

Instead, the Pragmatist contingent contends, environmental philosophers could better spend their time and intellectual capital helping lay people clarify their actual environmental values—as opposed to speculating about some newfangled value which they would then try to impose on lay people—and helping lay people sort out what to do in the context of specific problems or issues.[40] Often we may find that

[38] See, e.g., Light, "Compatibilism in Political Ecology";
Light, "Environmental Pragmatism" and "Environmental Ethics and Weak Anthropocentrism";
Norton, *Toward Unity Among Environmentalists,* "Epistemology and Environmental Values," and "Why I am Not a Nonanthropocentrist";
Weston, "Beyond Intrinsic Value" and "Before Environmental Ethics."

[39] Norton, *Toward Unity Among Environmentalists.*

[40] Light, "Environmental Pragmatism."

conflicting values support the same policy—as, for example, when those who value waterfowl for hunting and those who value it for watching can support waterfowl habitat preservation and restoration policies—and philosophers can help lay people figure that out.[41] This is characterized as a more bottom-up, rather than top-down, approach to environmental policy.[42] Begin with something specific and local, such as a scheme to develop a forested landscape or to dam a stream and create a lake, or a plan to rehabilitate an abandoned mine site, or to reintroduce an extirpated predator. The role of environmental philosophers in environmental policy and decision-making processes is to bring the tools of conceptual analysis, values clarification and, yes, ethical theory, to bear on the problem—but only to the extent that theory is familiar (and thus conventional), easily understandable and illuminating; and to the extent that the problem itself determines what theories are useful to its solution.

THE PRACTICAL EFFICACY OF THEORETICAL ENVIRONMENTAL PHILOSOPHY

I have no quarrel whatsoever with the bottom-up approach to environmental philosophy. I myself was a recipient of a three-year grant from the bi-national Great Lakes Fishery Commission to work with an ichthyologist and an aquatic community ecologist to re-envision fishery management policy in the Great Lakes for the new millennium. My role was precisely to clarify such fuzzy conservation concepts as biological integrity, ecosystem health, ecosystem management, ecological restoration, ecological rehabilitation, ecological sustainability, sustainable development, and adaptive management; and to examine the values that have driven, drive and will drive fishery management in the Great Lakes in the past, present and future.[43] I do have a quarrel, however, with the representation of the bottom-up, Pragmatic approach as a competitive alternative to theoretical environmental philosophy and to the invidious comparison that environmental Pragmatists make between the two, virtually insisting that theorists should stop their pointless and pernicious theorizing.[44] I believe that the two—theory and practice—should be complementary, not competitive. Further, I think that theoretical environmental philosophy is powerfully pragmatic: theory *does* make a difference to practice.

What difference? First, the convergence hypothesis—which Norton confesses is merely "an article of environmentalists' faith"[45]—is not a credible article of faith because it is hard to believe that all of Earth's myriad species, for example, are in

[41] Norton, *Toward Unity Among Environmentalists.*

[42] See, e.g., Norton, *Toward Unity Among Environmentalists;*
Weston, "Beyond Intrinsic Value" and "Before Environmental Ethics";
Minteer, B.A. "Intrinsic Value for Pragmatists?"

[43] Callicott, Crowder, and Mumford.

[44] Norton (1992, 1995), *Op. cit.*; Norton, "Epistemology and Environmental Values" and "Why I am Not a Nonanthropocentrist"; Minteer, "No Experience Necessary?".

[45] Norton, *Toward Unity Among Environmentalists,* 241.

some way useful to human beings.[46] Many may represent unexplored potential: new pharmaceuticals, foods, fibers and tools. But many more may not.[47] Many species that have no actual or potential resource value are critical agents in ecological processes and/or perform vital ecological functions or "services." But many more do not.[48] Many non-resource, non-ecological-service-provider species are, nevertheless, objects of aesthetic wonder and/or epistemic curiosity to the small percentage of the human population that is aesthetically cultured and scientifically educated. But such amenity values that endangered non-resource, non-ecological-service-provider species have for a tiny human minority affords them little protection in a world increasingly governed by market economics and majority-rule politics. In short, conservation policy based on anthropocentrism alone—however broadened to include potential as well as actual resources, ecosystem services, and the aesthetic, epistemic, and spiritual uses of nature by present *and future* people—is less robust and inclusive than conservation policy based on the intrinsic value of nature.[49]

Second, in setting forth the "convergence hypothesis," Norton focuses exclusively on the content of anthropocentric and nonanthropocentric (or intrinsic) values and the environmental policies they support.[50] But if we focus instead on the formalities, as it were, or structural features of the policy discourses involving, on the one hand, claims of intrinsic value; and those, on the other, that only involve anthropocentric value claims, a hypothesis contrary to the "convergence hypothesis" is suggested. Perhaps it should be called the "divergence hypothesis."

Broad recognition of the intrinsic value of human beings places the burden of proof on those who would override that value for the sake of realizing instrumental values. For example, an intrinsically valuable human being not wishing to sell a piece of property at any price may refuse any offer to buy it. Their intransigence, however, may be trumped if benefits to the public rise beyond a certain threshold. If, for example, the recalcitrant owner's property stands in the way of an urban light-rail track, then the property may be "condemned," and the owner paid fair market value for it, whether he or she is willing to sell it or not. If nature were also broadly recognized to have intrinsic value the burden of proof would shift, *mutatis mutandis,* from conservators of nature to exploiters of nature.[51] If something has only instrumental value, its disposition goes to the highest bidder. If that something is some subsection of nature—say, a wetland—conservationists must prove that an economic cost-benefit analysis unequivocally indicates that it has greater value as an amenity than it has, drained and filled, as a site for a proposed shopping mall. But if the intrinsic value of wetlands were broadly recognized, then developers would have to prove that the value to the human community of the shopping mall was so great as to trump the intrinsic value of the wetland. The concept of intrinsic value in nature

[46] Ehrenfeld, "The Conservation of Non-Resources" and "Why Put a Value on Biodiversity?"
[47] Ehrenfeld, "The Conservation of Non-Resources."
[48] Ehrenfeld, "Why Put a Value on Biodiversity?"
[49] Ehrenfeld, "The Conservation of Non-Resources" and "Why Put a Value on Biodiversity?"
[50] Norton, *Toward Unity Among Environmentalists.*
[51] Fox, 101.

functions politically much like the concept of human rights. Human rights—to liberty, even to life—may be overridden by considerations of public or aggregate utility. But in all such cases, the burden of proof for doing so rests not with the rights-holder, but with those who would override human rights. And the utilitarian threshold for overriding human rights is pitched very high indeed. As Fox puts it:

> The mere fact that moral agents must be able to justify their actions in regard to their treatment of entities that are intrinsically valuable means that recognizing the intrinsic value of the nonhuman world has a dramatic effect upon the framework of environmental debate and decision-making. If the nonhuman world is only considered to be instrumentally valuable then people are permitted to use and otherwise interfere with any aspect of it for whatever reasons they wish (i.e., no justification is required). If anyone objects to such interference then, within this framework of reference, the onus is clearly on the person who objects to justify why it is *more useful* to humans to leave that aspect out of the nonhuman world alone. If, however, the nonhuman world is considered to be *intrinsically valuable* then the onus shifts to the person who wants to interfere with it, to justify why they should be allowed to do so: anyone who wants to interfere with any entity that is intrinsically valuable is morally obliged to be able to offer a *sufficient justification* for their actions. Thus recognizing the intrinsic value of the nonhuman world shifts the *onus of justification* from the person who wants to protect the nonhuman world to the person who wants to interfere with it—and that, in itself, represents a fundamental shift in the terms of environmental debate and decision-making.[52]

THE PRAGMATIC POWER OF THE RIGHTS DISCOURSE

Mention of human rights leads to my third and last point about the pragmatic power and practical difference of theoretical environmental philosophy and its preoccupation with the concept of intrinsic value in nature. Human beings have shoes, teeth, kidneys, thoughts and rights. Human shoes and teeth are out there for anyone to see. Human kidneys may be observed during surgery or autopsy. We are privy only to our own thoughts and infer the thoughts of others from what they do, what they say, and what they write. However open to view or hidden away, human shoes, teeth, kidneys and thoughts are all things of this world. But "human rights" is a name for nothing: it is but an idea—a fiction—created by Western moral philosophers.[53] Theoretical moral philosophers created, more generally, a rights *discourse* in the West.[54]

When it was fresh and new, other moral philosophers tried to silence that discourse, for various reasons. For example, in the eighteenth century Jeremy Bentham, infamously, dismissed the idea that human beings have rights as "nonsense on stilts."[55] But the human-rights discourse survived its political and philosophical naysayers. It was institutionalised in the West by the adoption of the Bill of Rights, the first ten amendments to the Constitution of the United States, in 1789. It was globalized by the adoption of the Universal Declaration on Human

[52] Ibid., 101.
[53] Nickel.
[54] Gewirth.
[55] Ibid.

Rights by the United Nations General Assembly in 1948, now translated and published in 300 languages.[56] Presently, the United Nations International Bill of Human Rights consists of the Universal Declaration plus other human-rights measures adopted during the 1950s and 1960s—the International Covenant on Economic, Social and Cultural Rights, and the International Covenant on Civil and Political Rights and its two Optional Protocols, one on civil and political rights, one on the abolition of the death penalty.[57] The United Nations maintains an active (and geopolitically important) Commission on Human Rights and an office of "High Commissioner for Human Rights." Human-rights discourse, throughout the latter half of the twentieth century and the beginning of the twenty-first, has had enormous pragmatic effect worldwide as an instrument of criticism and political reform.[58] In the name of human rights, we condemn everything from "female circumcision" in parts of Muslim Africa to the Tienamen Square massacre in China, and reform of everything from the political status of American Indians in the United States to that of brides in India.

Especially in the subjectivist version that I endorse, the concept of intrinsic value of nature, like the concept of human rights, designates a less substantive thing than a pragmatic limit on policies driven by aggregate utility. Practically by definition, the *adjective* "intrinsic" entails that the character or property it modifies exists objectively in the entity to which it is attributed. Indeed, often the adjective "intrinsic" means that the character or property it modifies is the very essence of the entity to which it is attributed. For example, transporting oxygen to tissues in organisms is intrinsic to hemoglobin; competition is intrinsic to sport; volatility is intrinsic to the gaseous state of matter. In environmental philosophy, however, "intrinsic value" has also been consistently implicitly defined, *via negativa,* as the antonym of "instrumental value." What value remains—if any does—after all of something's instrumental value has been accounted for is its intrinsic value. Personally, I want to be useful to my family, friends, colleagues, neighbors, fellow citizens, and to my various human communities, and to the biotic community. But when the time comes (and regardless of whether it should come because of age, infirmity, or both) that I cease to be of instrumental value, I shall still value myself intrinsically and expect others to treat me that way as well (or at least treat me as if they did). Thus to value something intrinsically—as we shift from the adjectival-objective to the adverbial-subjective form—is to value something for itself, as an end-in-itself (to reinvoke the Kantian mode of expression), not merely as a means to our own ends, not merely as an instrument. From this perspective, there is no objective property in entities to which the noun "value" corresponds. Rather, we subjects value objects in one or both of at least two ways—instrumentally or intrinsically—between which there is no middle term.

[56] United Nations, "The International Bill of Human Rights."
[57] Ibid.
[58] Ibid.

THE PRAGMATIC EFFICACY OF THE INTRINSIC-VALUE-IN-NATURE DISCOURSE

Pragmatist philosophers now carp and cavil against the concept of intrinsic value in nature as still more nonsense on stilts. Bryan Norton, for one, has carried on a virtual jihad against the idea.[59] But environmental activists—for example, Dave Foreman, founder of Earth First!, the most radical group of environmental activists in the United States—have appreciated its practical efficacy. A while ago, Foreman wrote:

> Too often, philosophers are rendered impotent by their inability to act without analysing everything to absurd detail. To act, to trust your instincts, to go with the flow of natural forces, *is* an underlying philosophy. Talk is cheap. Action is clear.[60]

Later, Foreman changed his tune. He identified four forces that are shaping the conservation movement at the dawn of the new millennium. They are, and I quote, first "academic philosophy;" second, "conservation biology;" third, "independent local groups;" and fourth, "Earth First!"[61] That's right, "academic philosophy" heads the list. This is some of what Foreman has to say about it:

> During the 1970s, philosophy professors in Europe, North America, and Australia started looking at environmental ethics as a worthy focus for discussion and exploration… By 1980, enough interest had coalesced for an academic journal called *Environmental Ethics* to appear… An international network of specialists in environmental ethics developed, leading to one of the more vigorous debates in modern philosophy. At first, little of this big blow in the ivory towers drew the notice of working conservationists, but by the end of the '80s, few conservation group staff members or volunteer activists were unaware of the Deep Ecology—Shallow Environmentalism distinction or of the general discussion about ethics and ecology. At the heart of the discussion was the question of whether other species possessed intrinsic value or had value solely because of their use to humans [and]… what, if any, ethical obligations humans had to nature or other species.[62]

Notice that for the discourse of intrinsic value and, more generally, environmental ethics to have practical effect, it was not necessary for "working conservationists" to follow the ins and outs of the "big blow in the ivory towers." Such philosophical niceties as what properly justifies or grounds the intrinsic value of nature, which natural entities possess intrinsic value and which do not, and whether intrinsic value is an objectively existing supervenient property or is subjectively attributed and defined negatively as opposed to instrumental value, were not of the least importance. All that was important was that working conservationists were aware of the anthropocentric/nonanthropocentric distinction and the fact that there was a "general discussion about ethics and ecology" going on among environmental philosophers, "at the heart" of which "was the question of

[59] See, e.g., Norton, "Environmental Values and Weak Anthropocentrism," "Epistemology and Environmental Values," *Toward Unity Among Environmentalists,* and "Why I am Not a Nonanthropocentrist."

[60] Foreman, "More on Earth First!"

[61] Foreman, "The New Conservation Movement."

[62] Ibid., 8.

whether other species possessed intrinsic value or had value solely because of their use to humans." Note the parallel with human-rights discourse. Few human rights advocates and activists are conversant with the debate among moral philosophers about whether human rights are natural, God-given, or the wholly artificial product of a "social contract." It is the general idea under philosophical discussion that fires up the imaginations of lay people, morally inspires them, and reorients their perception of the world—the social world in the case of human rights, and the natural world in the case of nonanthropocentric environmental ethics.

The intrinsic-value-in-nature discourse soon spread from "conservation group staff members and volunteer activists" to professional natural resources managers. For example, in my work for the Great Lakes Fishery Commission, I found "intrinsic value"—by that name—attributed to the fishes of Lake Superior in a management plan produced by the Minnesota Department of Natural Resources. In a recent review of the philosophical debate about intrinsic value in nature, Christopher Preston points out the various domains of discourse that the concept of intrinsic value in nature has now penetrated.[63] In addition to that of environmental activists and government-agency environmental professionals, it crops up in the discourse of the new field of ecocriticism—in discussions of nature poets, such as William Wordsworth, Robinson Jeffers and Gary Snyder; and of nature writers, such as Edward Abby, Annie Dillard and Barry Lopez. According to Preston, the concept of intrinsic value in nature is "latent" in some U.S. environmental laws—the Wilderness Act of 1964, the 1973 Endangered Species Act, for example—and in some international declarations and treaties, such as the 1982 World Charter for Nature and the Global Biodiversity Treaty, signed by 160 countries (not including the United States) at the Earth Summit in Rio de Janeiro in 1992.

THE EARTH CHARTER: A UNIVERSAL DECLARATION OF INTRINSIC VALUE IN NATURE

Preston concludes that "[t]here is plenty of evidence to suggest that belief in intrinsic value in nature is playing an increasingly prominent role in the formation of environmental attitudes *and policies* worldwide."[64] One might protest that that depends on what is meant by "worldwide." If Preston means that the concept is pragmatically efficacious worldwide because belief in intrinsic value in nature is playing an increasingly prominent role in the formation of environmental attitudes and policies in North America, Western Europe, Australia, and New Zealand, he is surely correct. But if he means also to suggest that it is pragmatically efficacious in such countries as China and India, the world's two largest, some may have reason to doubt his claim. I have no experience in India, nor in the People's Republic of China, but I have been invited to lecture extensively in the Republic of China (Taiwan) and so can say from personal experience that many Taiwanese

[63] Preston.
[64] Ibid., 411, emphasis added.

environmental NGOs partially cast their activities in the discourse of the intrinsic value of nature. As to India, the evidence is contradictory. Ramachandra Guha, in a justly famous article, argued that although

> the transition from an anthropocentric (human-centered) to a biocentric (humans as only one element in the ecosystem) view in both religious and scientific traditions is to be welcomed,... this dichotomy is, however, of little use in understanding the dynamics of environmental degradation.[65]

By implication, presumably, it would be of little use in opposing the dynamics of environmental degradation. Vandana Shiva, on the other hand, in a justly famous book, argues that in popular traditional Indian belief, nature is an active subject—not a passive object, as in modern Western thought.[66] Neither Guha nor Shiva focus their discussions specifically on the concept of intrinsic value in nature, though Guha's somewhat equivocal discussion of "biocentrism" and Shiva's approving discussion of nature as active subject could, by implication, be understood as bearing on it.

There is another piece of evidence supporting the worldwide currency of the concept of intrinsic value of nature not mentioned by Preston that is much less problematic. After more than a decade of worldwide "consultations" with thousands of people representing millions of constituents in hundreds of interest groups and political-identity groups, the Earth Charter Commission issued a final draft of an "Earth Charter" in March, 2000. The idea of an Earth Charter was first conceived during preparations for the 1992 United Nations Conference on Environment and Development (a.k.a. the Earth Summit). Afterward, the Commission was formed to draft a document that would be circulated throughout the world for comment and revision, finally to be submitted to the United Nations for endorsement by the General Assembly in 2002, on the tenth anniversary of the Earth Summit. The very first principle of the Earth Charter reads: "1. Respect Earth and life in all its diversity. a) Recognize that all beings are interdependent and *every form of life has value regardless of its worth to human beings.*"

The *phrase* "intrinsic value" does not appear in the final draft of the Charter—although it did in preliminary drafts, including the penultimate one. The *concept* seems to remain, however, in the statement that "every form of life has value regardless of its worth to human beings." A diehard environmental Pragmatist opposed to the concept of intrinsic value in nature and determined to suppress it could argue that these words of the Earth Charter should be interpreted to mean that every form of life may have instrumental value for forms of life other than human beings, but such would be a tortured interpretation. Such an interpretation implicitly assumes, moreover, that if not "every" then some nonhuman forms of life have intrinsic value—else why must we care about what is of instrumental value to them? Further, arguments such as those of Ehrenfeld that many forms of life, often those most at risk of extinction, are "non-resources"—whether for humans or other kinds of being—implies that, as a matter of fact, not every form of life has instrumental

[65] Guha, 74.
[66] Shiva.

value.[67] In any case, the principal architect of the Earth Charter provides decisive comments on the proper interpretation of the words in question in response to my inquiry about the absence of the phrase "intrinsic value" in the final version of the document after it had appeared in all the previous ones:

> In your letter you express some concern about what may have been the anthropocentric orientation of some of our constituencies. You also identify the critical points in the text [those just quoted] where the ecocentric orientation of the Charter is made explicit. Throughout the document you will find that we have made a consistent effort to make clear that the moral community to which human beings belong extends beyond the human family to include the entire larger living world. In line with this outlook, the first principle, from which all the others flow, affirms respect for "Earth in all its diversity."[68]

I think that if the Earth Charter is eventually endorsed by the United Nations General Assembly, the result may well be comparable to the adoption of the Universal Declaration of Human Rights by the same body in 1948. The U.N. Universal Declaration of Human Rights was not a binding law or international treaty. But it did put the concept of human rights into play on the world stage. In effect, it globally institutionalised the discourse of human rights. Similarly, the Earth Charter may institutionalise and globalize the discourse of environmental ethics with its most potent concept of the intrinsic value of nature. In comparison with this achievement of theoretical environmental philosophers—the creation and dissemination of such a transformative discourse—the program of bottom-up environmental ethics recommended by Pragmatists appears quite modest and unambitious. Certainly, the energy and intellectual capital of theoretical environmental philosophy should not be redirected into such yeoman (and yeowoman) work; on the contrary it should be redoubled.

THE ENVIRONMENTAL PRAGMATIST CAPITULATION TO THE CONCEPT OF INTRINSIC VALUE IN NATURE

Minteer has recently argued that Pragmatists need not eschew the concept of intrinsic value in nature after all.[69] He even demonstrates that Norton himself, the most ardent opponent of the concept of intrinsic value of nature, actually endorses it, although Norton still refuses to use the term "intrinsic value" because it is "tainted."[70] This is ironic because the only reason it may seem to them to be tainted is that environmental Pragmatists, and especially Norton, have conflated the various not-so-subtly differing theories of intrinsic value (reviewed here) into one grotesque caricature, in their zeal to stamp out intrinsic-value theorizing in environmental philosophy. Minteer, who apparently relies on Norton to characterize (that is,

[67] Ehrenfeld, "The Conservation of Non-Resources" and "Why Put a Value on Biodiversity?"
[68] Rockefeller.
[69] Minteer, "Intrinsic Value for Pragmatists?"
[70] Ibid., 66.

caricature) intrinsic value theory in nature thinks that Rolston and I hold more or less the same theory while, as clearly noted here, our theories differ dramatically (my attempt to argue that Rolston is a crypto-subjectivist notwithstanding). Minteer insists, for example, on "the universalist and foundationalist uses of the concept by such theorists as Callicott and Rolston."[71] According to Minteer, Norton finds us guilty of "disengaged ontological and metaphysical solutions for environmental quandaries" and "of abstraction and ideological dogmatism among other vices."[72] Among these other vices are "foundationalism," "Cartesianism," and being "universalistic," "monistic," and even "intellectualistic." Minteer never explains just what these vices amount to, however. What, for example, is foundationalism? What does it mean to offer "ontological and metaphysical solutions for environmental quandaries" and why is this a vice? All Minteer does is sling these words around and rhetorically condemn them as "vices." Further, I have repeatedly tried to explain in what sense I advocate monism in environmental ethics and in what sense I do not.[73] All such niceties, however, are simply ignored by Minteer, who, despite my express declaration to the contrary, insists that I am a "reductionist" on a "quest for a universal master principle."[74]

Had Minteer bothered to read what Rolston, and especially what I, have actually written about intrinsic value in nature, rather than to rely on Norton's polemically inflamed caricature, he might have discovered that the kind of Pragmatist theory of intrinsic value that he recommends and seems to believe that he is articulating for the first time is, more or less, the same as I have long espoused. He writes,

> I do think we can, as pragmatists, accommodate noninstrumental values in our justifications of environmental policy. [W]e may value nonhuman nature noninstrumentally.[75]

And he insists that "humans 'do' the valuing, which may or may not be instrumental." That is pretty much what I have been arguing all along, with the proviso that lots of other forms of life can also "do" a bit of valuing. All value, in short, is of subjective provenance. And I hold that intrinsic value should be defined negatively, in contradistinction to instrumental value, as the value of something that is left over when all its instrumental value has been subtracted. In other words, "intrinsic value" and "noninstrumental value" are two names for the same thing.

Minteer frankly acknowledges that

[71] Ibid., 61.
[72] Ibid., 65.
[73] Callicott, *Beyond the Land Ethic.*
[74] Minteer, "Intrinsic Value for Pragmatists?", 65.
[75] Ibid., 72.

> we pragmatists have tended to neglect and often besmirch the worth and validity of intrinsic value claims in our enthusiastic embrace of a wide and deep instrumentalism, even if the former resonate with large segments of the public.[76]

He even goes so far as to acknowledge that "intrinsic value arguments might be the most powerful and effective in certain circumstances." But he also claims that intrinsic-value-in-nature theorists "disparage instrumental values." This is certainly not true. For example, I have written the following topic sentence for a chapter in a textbook on conservation biology and then have gone on to fully and sympathetically flesh out each topic: "The anthropocentric instrumental (or utilitarian) value of biodiversity may be divided into three basic categories—goods, services and information."[77] Because some conservation biologists have confused it with intrinsic value, I go on to cautiously note that "[t]he psycho-spiritual value of biodiversity is possibly a fourth kind of anthropocentric utilitarian value," which I also then fully flesh out.[78] The taxonomy of "value in nature" that Rolston develops is even more elaborate. In the abstract of one article he lists "(1) economic value, (2) life support value, (3) recreational value, (4) scientific value, (5) aesthetic value, (6) life value, (7) diversity and unity values, (8) stability and spontaneity values, (9) dialectical value, and (10) sacramental value."[79] In the abstract of another he "itemize[s] twelve types of value carried by wildlands[:] economic, life support, recreational, scientific, genetic diversity, aesthetic, cultural symbolization, historical, character building, therapeutic, religious and intrinsic."[80] Rolston thinks that appeal to *all* of them—and all but one are anthropocentric/ instrumental—by those wishing to preserve wildlands is both effective and legitimate. Thus it is anti-intrinsic-value-in-nature Pragmatists, not us more inclusive pro-intrinsic-value-in-nature theorists, who are "locking out those citizens from the moral debate who choose to speak about the value of nature in ways that" Norton and other Pragmatists "can[not] philosophically abide."[81]

Minteer is no more specific about Norton's arguments against intrinsic value in nature than he is about the nature of foundationalism or universalism. He vaguely refers to "the epistemic problems regarding the justification of intrinsic values as well as the metaphysical status of noninstrumental claims" discussed by Norton, but he provides no summary.[82] What Minteer does provide, however, is insight into Norton's motives. Norton is "primarily motivated by his desire to speak clearly and effectively to practical matters of environmental management and problem-solving."[83] Thus he "concluded early on," according to Minteer, that intrinsic value theory was a pragmatic "dead end and that a weak anthropocentric approach and a

[76] Ibid., 61.
[77] Callicott, "Conservation Values and Ethics," 29.
[78] Ibid., 30.
[79] Rolston, "Values in Nature," 113-128.
[80] Rolston, "Valuing Wildlands," 23-48.
[81] Minteer, "Intrinsic Value for Pragmatists?", 61.
[82] Ibid.
[83] Ibid., 62.

broad instrumentalism could deliver the goods."[84] But intrinsic value theory just would not go away as Norton wished it would. I am grateful to Minteer for documenting that Norton, despite his campaign against the concept of intrinsic value in nature, occasionally forgets himself and acknowledges its pragmatic power in supporting what Minteer calls "good environmental policy."[85] More importantly, I am also grateful to Minteer for candidly acknowledging what Norton has persistently denied—that the notion that nature has noninstrumental value is increasingly part of "the public's everyday intuitions and sentiments regarding nonhuman nature."[86] My main point in this essay is that the public might not have so commonly now valued nature noninstrumentally had the work of environmental philosophers not created a new discourse—the discourse of intrinsic value in nature, a new, positive and inspiring name, as opposed to the essentially privative term "noninstrumental"—in which the public's everyday intuitions and sentiments regarding nonhuman nature might be powerfully articulated.

CONCLUSION

We sometimes forget, I think, that we live, move and have our human being in a world of words, as well as in a physical world beyond words. For all its importance —which above all environmental philosophy affirms and celebrates—that world beyond human words is only accessible through the portal of human discourse. In conclusion, therefore, we must agree with Confucius that the first order of business is any policy arena is to rectify names, so that our policies and practices are framed in terms of the most efficacious and transformative discourse. The way Confucius would rectify names is by administrative fiat. In a democracy we do so by means of the free and sometimes technical philosophical discussion of frequently controversial and sometimes new and radical ideas. That discussion, especially if it is carried on largely in the academy, may seem far removed from the fray of public policy debate and hopelessly impractical. Yet in multiple and diffuse ways it seeps out of the ivory tower into the public domain, and finally funds the formation of public policy and practice. That has, demonstrably, been the case with theoretical environmental ethics and its central idea, the intrinsic value of nature.

[84] Ibid.
[85] Ibid., 71.
[86] Ibid., 60.

REFERENCES

Callicott, J.B. *Beyond the Land Ethic: More Essays in Environmental Philosophy.* Albany: State University of New York Press, 1999.

_______. "Conservation Values and Ethics." In *Principles of Conservation Biology,* 2nd edition. Edited by G.K. Meffe and C.R. Carroll, 22-35. Sunderland: Sinauer Associates, 1997.

_______. "The Pragmatic Power and Promise of Theoretical Environmental Ethics." *Environmental Values* 11 (2002): 3-25.

Callicott, J.B., L.B. Crowder, and K. Mumford. "Current Normative Concepts in Conservation." *Conservation Biology* 13 (1999): 22-35.

Descartes, R. *Discourse on Method.* New York: Liberal Arts Press, 1987. (*Discours de la méthod.* Paris: Michaelem Soly, 1637).

Ehrenfeld, D. "The Conservation of Non-Resources." *American Scientist* 64 (1976): 647-655.

_______. "Why Put a Value on Biodiversity?" In *Biodiversity.* Edited by E.O. Wilson, 212-216. Washington: National Academy Press, 1988.

Foreman, D. "More on Earth First! and *The Monkey Wrench Gang*." *Environmental Ethics* 5 (1983): 95-96.

_______. "The New Conservation Movement." *Wild Earth* 1, no. 2 (1991): 6-12.

Fox, W. *Toward a Transpersonal Ecology.* Boston: Shambhala Publications, 1990.

_______. "What Does the Recognition of Intrinsic Value Entail?" *Trumpeter* 10 (1993): 101.

Gewirth, A. "Rights." In *Encyclopedia of Ethics,* vol. 1. Edited by L.C. Becker and C.B. Becker, 1103-1109. New York: Garland Publishing, 1992 (2nd edition, 2001).

Ghiselin, M.T. "A Radical Solution to the Species Problem." *Systematic Zoology* 23 (1974): 536-544.

Goodpaster, K.E. "On Being Morally Considerable." *Journal of Philosophy* 75 (1978): 308-325.

Guha, Ramachandra. "Radical American Environmentalism and Wilderness Preservation: A Third World Critique." *Environmental Ethics* 11 (1989): 71-83.

Hall, D.L. and R.T. Ames. *Thinking Through Confucius.* Albany: State University of New York, 1987.

Hull, D. "Are Species Really Individuals?" *Systematic Zoology* 25 (1976): 174-191.

Johnson, L.E. *A Morally Deep World: An Essay on Moral Significance and Environmental Ethics.* Cambridge: Cambridge University Press, 1991.

Kant, I. *Foundations of the Metaphysics of Morals.* Translated by L.W. Beck. New York: Library of Liberal Arts, 1959 (*Grundlegung zur metaphysic der Sitten*. Leipzig: Felix Meiner, 1785).

Leopold, A. *A Sand County Almanac, and Sketches Here and There.* New York: Oxford University Press, 1949.

Light, A. "Compatibilism in Political Ecology." In *Environmental Pragmatism*. Edited by A. Light and E. Katz, 161-184. New York: Routledge, 1996.

________. "Environmental Pragmatism as Philosophy or Metaphilosophy?" In *Environmental Pragmatism*. Edited by A. Light and E. Katz, 325-338. New York: Routledge, 1996.

McIntosh, R.P. *The Background of Ecology: Concept and Theory.* Cambridge: Cambridge University Press, 1985.

Minteer, B.A. "Intrinsic Value for Pragmatists?" *Environmental Ethics* 23 (2001).

________. "No Experience Necessary?: Foundationalism and the Retreat from Culture in Environmental Ethics." *Environmental Values* 7 (1998): 333-347.

Moore, G.E. *Principia Ethica.* Cambridge: Cambridge University Press, 1903.

Naess, A. "The Shallow and the Deep, Long-Range Ecology Movements: A Summary." *Inquiry* 16 (1973): 95-100.

Nickel, J.W. "Human Rights." In *Encyclopedia of Ethics,* vol. 1. Edited by L.C. Becker and C.B. Becker, 561-565. New York: Garland Publishing, 1992 (2nd edition, 2001).

Norton, B.G. "Environmental Ethics and Weak Anthropocentrism." *Environmental Ethics* 6 (1984): 131-148.

________. "Epistemology and Environmental Values." *The Monist* 75 (1992): 208-226.

________. *Toward Unity Among Environmentalists.* New York: Oxford University Press, 1991.

________. "Why I am Not a Nonanthropocentrist: Callicott and the Failure of Monistic Inherentism." *Environmental Ethics* 17 (1995): 341-358.

O'Neill, J. "Varieties of Intrinsic Value." *The Monist* 75 (1992): 119-137. Routley, R. and V. Routley. "Human Chauvinism and Environmental Ethics." In *Environmental Philosophy.* Edited by D. Mannison, M. McRobbie, and R. Routley, 96-189. Canberra: Department of Philosophy, Research School of the Social Sciences, Australian National University, 1980.

Preston, C. "Epistemology and Intrinsic Values: Norton and Callicott's Critiques of Rolston." *Environmental Ethics* 20 (1998): 409-428.

Regan, T. "An Examination and Defense of One Argument Concerning Animal Rights." *Inquiry* 22 (1979): 189-219.

________. *The Case for Animal Rights.* Berkeley: University of California Press, 1983.

________. "The Nature and Possibility of an Environmental Ethic." *Environmental Ethics* 3 (1982): 19-34.

Rockefeller, S.C. Letter to author. July 12, 2000.

Rolston III, H. *Conserving Natural Value.* New York: Columbia University Press, 1994.

________. *Environmental Ethics: Duties to and Values in the Natural World.* Philadelphia: Temple University Press, 1988.

________. "Is There an Ecological Ethic?" *Ethics* 85 (1975): 93-109.

________. "Naturalizing Callicott." In *Land, Value, Community: Callicott and Environmental Philosophy.* Edited by W. Ouderkirk and J. Hill, 107-122. Albany: State University of New York Press, 2002.

________. "Values in Nature." *Environmental Ethics* 3 (1981): 113-128.

________. "Valuing Wildlands." *Environmental Ethics* 7 (1985): 23-48.

Savage-Rumbaugh, S., S.G. Shanker, and T.J. Taylor. *Kanzi: The Ape at the Brink of the Human Mind.* New York: Oxford University Press, 1998.

Shiva, V. *Staying Alive: Women, Ecology and Development.* London: Zed Books, 1988.

Singer, P. *Animal Liberation: A New Ethics for Our Treatment of Animals.* New York: Avon, 1977.

Sylvan, R. "Is There a Need for a New, an Environmental Ethic?" In *Environmental Philosophy*, 3rd edition. Edited by M.E. Zimmerman, J.B. Callicott, G. Sessions, K.J. Warren, and J. Clark. Upper Saddle River: Prentice-Hall, 2001.

Taylor, P.W. *Respect for Nature: A Theory of Environmental Ethics*. Princeton: Princeton University Press, 1986.

United Nations. Fact Sheet No. 2 (Rev. 1). *The International Bill of Human Rights*. Geneva: United Nations, 1996.

Weston, A. "Before Environmental Ethics." *Environmental Ethics* 14 (1992): 321-338.

________. "Beyond Intrinsic Value: Pragmatism in Environmental Ethics." *Environmental Ethics* 7 (1985): 321-339.

CHALMERS CLARK

THE EXPANDING CIRCLE AND MORAL COMMUNITY—NATURALLY SPEAKING[1]

1. INTRINSIC VALUE AND MORAL COMMUNITY

Much literature has been directed at criticizing the shortcomings of the anthropocentric paradigm in environmental philosophy. In this volume, J. Baird Callicott surveys the ways in which environmental philosophers have considered the question of intrinsic value as they seek to motivate an environmental policy that distributes moral worth beyond the confines of the species *Homo sapiens*.

Callicott goes on to respond to accusations levelled by the so-called "pragmatist" movement in environmental philosophy. Representatives of that movement contend that "environmental ethics has made little impact on environmental policy because environmental ethics has been absorbed with arcane theoretical controversies, mostly centered on the question of intrinsic value in nature." Holmes Rolston III and Callicott himself have been targeted under such critiques charging, among other things, that they are guilty of "disengaged ontological and metaphysical solutions for environmental quandaries" and "of abstraction and ideological dogmatism among other vices." Callicott is right to claim that such charges are based upon a serious misreading of his views, and his endorsement of the "pragmatic power and promise of theoretical environmental ethics" seems to hold.

In this essay I would like to sketch a way to consider moral concern for the environment that, to some degree, obviates the question of "disengaged ontological and metaphysical solutions" that environmental pragmatists so eschew. This inquiry will move us *beyond* the anthropocentric circle, but it may also be amenable to those who staunchly maintain an anthropocentric environmental ethic. My effort in this direction will be explicitly naturalistic rather than metaphysical. I will do this in several phases. First, I will discuss the limitations of anthropocentrism as articulated by Peter Singer. Second, I will offer a prompt from Plato that poses the question of moral action and self-interest. Third, I will use Quine's naturalistic reasoning to discuss how care for others may be part and parcel with our own self-interests. I will then tie up these individual strands of argument, and I will suggest a notion of stewardship in which *trust* figures prominently. This qualified sense of stewardship may be one way to enact our commitments to the intrinsic value of other species or natural entities.

[1] This essay was written at Yale University's Interdisciplinary Bioethics Project with the support of the Donaghue Initiative in Biomedical and Behavioral Research Ethics.

A. W. Galston and C. Z. Peppard (eds.), Expanding Horizons in Bioethics, 209-220.

1.1 A Note on Naturalism

The naturalism of which I speak derives from the naturalized epistemology of W.V. Quine, perhaps the most influential Anglo-American philosopher of the last fifty years. Simon Blackburn speaks of Quine's naturalistic approach as an "enterprise of studying the actual formation of knowledge by human beings…" Since Quine is concerned with human knowledge and learning, his philosophy tends to "blend into the psychology of learning and the study of episodes in the history of science."[2] How Quine's naturalism relates to environmentalism will be shown below.

1.2 Peter Singer, "Speciesism," and Implications for the Moral Community

In this essay I will use Singer's influential critique of anthropocentrism, because his concept of "speciesism" has been especially indicative of the assault on a narrow conception of moral worth that assigns moral consideration to all, and only, members of the species *Homo sapiens*. Singer has noted that a key element from his argument was actually articulated about 125 years before he published *Animal Liberation.*[3] At a feminist convention in the 1850s, black feminist Sojouner Truth responded to the common view that blacks were less intelligent than whites by asking a more fundamental question: "What's [intelligence] got to do with women's rights or Negroes' rights? If my cup won't hold but a pint and yours holds a quart, wouldn't you be mean not to let me have my little half-measure?"[4] The issue Ms. Truth identifies might be called "a principle of equal consideration" in light of one's capacities, rather than a call for equal treatment irrespective of them. Explicitly, the point is that moral rights do not depend upon superior or inferior degrees of a capacity—such as intelligence—but they do require equal consideration for one's *interests* in light of those capacities. If we apply the principle of equal consideration to the case of intelligence generally, it would turn out that while a child with Downs Syndrome should not be *treated as equal* to other normal children (i.e., *identical treatment*), the child should nonetheless be *given equal consideration* with respect to her interests, irrespective of how the child's capacities measure up across human lines. It was with this crucial insight that Singer then challenged the idea of limiting equal consideration of interests within a single species, i.e. *Homo sapiens*.

Much in the challenge to the anthropocentric perspective seems to hinge on the conception of interests, so let us pause for a moment to say a few words about the meaning of the term "interests." Roughly speaking, the interests of a being are those things one needs to flourish and to live a good life according to the extent of one's capacities. Singer mentions the contrast of kicking a stone down a path to amuse oneself and kicking a mouse. In the former case, he says, the stone clearly has no interests being violated, but the mouse, in fact, does have an interest not to be subject to the suffering caused by such actions. Here, Singer motivates his

[2] Blackburn, 255.

[3] Singer, *Animal Liberation.*

[4] Singer, "All Animals are Equal," 156.

understanding of "interests" in terms reaching back to Bentham. Bentham suggests that in tracing "the insuperable line" of moral considerability for any being, "the question is not, Can they reason? nor Can they *talk?* but, *Can they suffer?*"[5]

The argument regarding equal consideration of interests thus directs us to consider a close analogy between the question of moral consideration across species boundaries and questions of racism, sexism, and the like. In the face of the argument about equal moral consideration, Singer urges that to persist with an anthropocentric view of moral consideration is to fall victim to what he terms "speciesism." Speciesism, Singer says, is "a prejudice or attitude of bias toward the interests of members of one's own species and against members of other species."[6]

These arguments expose the arbitrary structure of anthropocentrism, even though it may be—at first glimpse—understandable that most of us are drawn to aid or to relate to those most like ourselves. Nonetheless, on the weight of the above logic, it becomes clear that the moral community—the range of beings with intrinsic value and thus a claim to moral consideration—must be cast significantly beyond the confines of the anthropocentric paradigm. However, Singer's argument amounts only to the negative claim that the species *Homo sapiens* cannot be viewed as the sole domain of intrinsic value and moral consideration. The question of what to include in the expanding circle of moral consideration now obtrudes.

2. THE EXPANDING CIRCLE: WHY CARE?

Accepting that the anthropocentric paradigm has significant problems, we might worry about how to extend the question of moral consideration across species boundaries. How far should our moral consideration extend, and which species or entities should get priority? Even if we accept the nonanthropocentric paradigm, we should ask two questions. First, are there philosophical grounds on which to expand the circle of moral consideration? (In this essay, I have employed Singer's argument to affirm that there are such philosophical grounds.) Second, are there reasoned grounds within this view that could motivate a more responsible environmental policy?

In an odd way, this concern leads us to a philosophical problem powerfully articulated in Plato's *Republic*: Why be moral in the first place? This problem is posed to Socrates by Thrasymachus, who argues that that justice, or morality, is simply that which is in the interests of the stronger. Socrates goes on to philosophically disable Thrasymachus by teasing out a contradiction in his account, but Glaucon remains unconvinced of the general defeat of the main claim made by Thrasymachus. Eager to see its complete demise, Glaucon challenges Socrates to address Thrasymachus's point anew as he offers the myth of the ring of Gyges. Gyges, it seems, discovered a ring that would allow him to disappear anytime he

[5] One can also draw a distinction between what might be called our real interests in opposition to our apparent interests. For example, even if a child does not desire a college education, it is in the child's "real" interest to acquire an education if the child has the capacity for it.

[6] Singer, "All Animals are Equal," 157.

wished. Crucial to the myth is an intuition that if we had a magic ring that could allow us to disappear and reappear at our pleasure—like the ring found by Gyges—such a finding would drastically weaken our interest in being moral in the first place. With possession of the ring we could take what we want and mystify or evade those who might try to pursue or stop us. It's a soul searcher, and a slippery slope: If we had Gyges' ring, would we eventually abandon what we normally call morality altogether?

The symbolism of the ring is offered to suggest that morality is practiced only because we lack sufficient power to gain all that we would otherwise seize on grounds of self-interest. Morality, from this point of view, is merely a prudential thing, an acknowledgement of our own limitations. In this scheme, morality becomes a compromise meant to maximize our own self-interests. It is framed against a complex mix of the competing powers and interests of others, all of which can deny or hinder us in achieving our desires. The upshot is that morality, as such, is not its own reward. It is a stop-gap measure, practiced prudentially, reluctantly, and against the grain of our natural human inclinations.[7] From this perspective, it seems unlikely that we would grant moral consideration to nonhuman species based on their intrinsic value. Nor would we enact environmental policy in accord with those moral considerations, because human dominion over the natural world is so extensive. If might makes right, environmental policy based on anything but humans' own desires is a very tough goal indeed. As such, from the perspective of Thrasymachus, it would follow that the interests of other species, if there are such, turn out to be matters of blithe indifference as long as we face no significant challenge to our own interests. Morality, on this account, is simply the interests of the stronger, and it is we who are the stronger.

3. CARING ABOUT HOW WE CARE: A NATURALISTIC-ARISTOTELIAN ARGUMENT

To motivate a basis to move beyond the limitations of the anthropocentric paradigm, I will build upon some broad naturalistic commitments. That is, I will ground my claims in a more empirical picture of the natural world, rather than in abstract metaphysical arguments. I will start with some basic features of the human species. The first naturalistic point is one that has gained considerable strength in both the biological sciences and the philosophy of biology. It is the thesis that there is a biological grounding for altruism in evolutionary theory. Kin selection, in particular, presents a case for seeing a genetic disposition towards altruism. Quine has commented upon this literature as follows:

> We must simply recognize that there are drives other than self-interest, and admirable ones. Ethologists represent some altruistic drives as innate in man and other animals, and they explain them by natural selection as ways of safeguarding the gene pool through the protection of kin. But man's altruism is not always as abundant as we could wish, nor are arguments from self-interest the way to increase it. The way rather is to

[7] Plato, 357a-369.

> play on whatever faint rudiments of fellow-feeling he may be capable of, fanning any little spark into a perceptible flame. Try the formative years for best results.[8]

While Quine suggests that arguments from self-interest are not the way to encourage altruism, I think it is important to recognize that if such altruistic drives are written into our genetic make-up, then there is an important avenue of self-interest that might be satisfied by acting *in accord with our drives* to promote the interests of others, rather than thwarting such drives.

Of course, acting to promote such forms of satisfaction would need to be qualified according to Aristotelian moral injunctions that such things be done in the right way, at the right time, for the right reason, and so on. One might even call this naturalistic picture a "compatibilist" position between psychological egoism and altruism. The view, quite simply, is that human beings have a mix of both self-centered and other-centered drives. Such compatibilism in drives, however, might also be seen as a support for *theoretical* altruism. No clear-headed altruist would deny that a great many human drives are self-interested. It is only the psychological egoist who, in the spirit of Hobbes, tells us we cannot be *both* altruistic and egoistic. Thus, we must avoid two fallacies: first, the fallacy that self-interest is selfishness (going to the dentist is self-interested but certainly not selfish); and second, the fallacy of thinking that because a drive comes *from* the self it is always *for* the self. If we avoid these two subtle fallacies, the way is open to view some form of altruism as basic. Further, we can say that it is not necessarily in conflict with our own self-interest.

I would like to draw another distinction from Quine's way of putting the matter in an effort to block a false assumption. Namely, we should not assume that since a genetic disposition towards altruism is inherited, the disposition is limited to only those who share our inherited genotype. Quine writes:

> Evolutionary theory accounts for innate altruism only toward kin. There is no mistaking the grading off of altruistic impulse as we move outward from kin; we are less protective of community than of family and still less of nation or race.... Philanthropy is girdling the globe and even reaching out to subhuman species. Dilemmas arise between human welfare and the welfare of other mammals. The human heart is distended until it is as big as all outdoors.[9]

In the final sentence of this passage Quine clearly sees the possibility of extending care for others, broadly. "The human heart," he says, may be "distended until it is as big as all outdoors." However, one might object that this doesn't follow from gene-based altruism, or what some mistakenly call the ways and means of the "selfish gene." On this limited view of altruism, human DNA is only "concerned" with preserving its own structure, its own genotype, and its own genotypic kin.

But this limited view is mistaken. First of all, the chemical structure of a nucleic acid that determines a particular protein is not "selfish" in the way that we understand human beings to be "selfish." Second, any sense of innate altruism "only towards kin" needs to be considered as a mix of phenotypic strategies for kin recognition, rather than tightly tethered forms of tracking genotypic identity. For

[8] Quine, 5.
[9] Ibid.

example, a well-documented phenotypic strategy for parental kin recognition in certain ducks is for the ducklings to follow the first large moving object it sees and imprint upon that object. Such a strategy has a high probability of succeeding to promote kin recognition in the natural environment of the ducks, but it has also allowed researchers to "fool" ducklings into imprinting upon and following the researchers. As such, recognition of kin by gross phenotypic expressions can float free of genetic identity of kin altogether. Indeed, I would like to claim that our altruistic attachments can float free species genotype as well. This outcome has substantial repercussions for the possibility of cross-species altruism. It is a naturalistic way to make some initial sense of Quine's expression that the human heart may be "distended…[to] all out doors."

The naturalistic arguments above imply that we are inclined to care about others, at least to some extent; and, further, that it is by dint of the forces of natural selection that we do so. Indeed, it appears that we are "condemned to care" by nature herself, although that "caring" often—though not always—benefits our gene pool. However, as Quine also notes, "man's altruism is not always as abundant as we could wish," for "there is no mistaking the grading off of altruistic impulse as we move outward from kin." So the question arises anew: Why seek to distribute intrinsic value beyond a narrow sphere of kin? Why be more moral than necessary? To answer this question, I want to bring the matter back to dimensions of self-interest rather than altruism as such. The motive harkens back to Aristotle and his quest for the good life (*eudaimonism*). It might be rendered thus: If we *must* do something, we do better if we do it well. If we can use reason (influenced by our surroundings and social environment) to guide and express our "innate" genetic faculties, then it also increases our well-being. This essay proposes, in relation to environmental ethics and policy, that we should do just that.

So, back to Glaucon's question, this time leveled across species lines: If we could get away with caring only for ourselves, would we? Why be moral in the first place? Our naturalistic answer is that since we are naturally disposed to care for others, then it is in our interests to do it wisely and well. It is probably correct to think that the cause of our caring is grounded in the genetics of kin selection; it is harder to claim that the *objects*, and the terms, of our caring are predicated only on genetic similarity. It would be a peculiar version of the genetic fallacy to think that we can only care about that to which we are genetically related. Indeed, as Quine remarked, the "human heart may be distended to include all outdoors." One can see the same operations at work within a variety of human families across genetic lines. If a child is unknowingly switched at birth, or if a child is adopted, the care that develops for the child is clearly based on phenotypic interactions. Genetic kin recognition largely drops out of the question. We construct relations, in addition to being born into them.

4. POLICY IMPLICATIONS: NOBLESSE OBLIGE AND STEWARDSHIP

We have noted the difficulty of setting guidelines for action toward others—i.e., policy—if our morality is merely one that is concerned with the interests of the stronger. Indeed, with the exception of the world of microbes, human dominion over the natural world is largely unchallenged. Does this prove Thrasymachus right? Not necessarily. I have tried to make two significant claims in the preceding sections. First, it is illusory to think that there is a principled divide between species boundaries (following Singer); and second that naturalistically speaking, our interests include an element of caring for others on genetic levels that are also shaped by social constructs and attachments.

If we thus break the anthropocentric paradigm of moral considerability, we now face the problem of how far to extend the moral community. We also have a self-interested directive to do it well and wisely. The actual extent to which we ought to distribute moral consideration is a very large question that will not be taken up here. Instead, my goal in this essay is to offer a naturalistic challenge to the anthropocentric paradigm, rather than "disengaged ontological and metaphysical solutions for environmental quandaries." Indeed, I will articulate a naturalistically-based claim that supports those who propose a widely ranging distribution of moral consideration based on intrinsic value. This is the case whether moral consideration is extended to individuals of various non-human species, to species themselves, or to something more like a holistic land ethic that "changes the role of *Homo sapiens* from conqueror of the land-community to plain member and citizen of it."[10]

These efforts are preliminary, and as such there are shortcomings. Quine has seen the way open to extend moral consideration to "all outdoors," but even if kin selection as the source of altruism isn't limited to genotype, "there is no mistaking the grading off of altruistic impulses as we move outward ..." In short, while we have opened the door, we haven't yet gotten very far from it. Nonetheless, there is a powerful sense that it is morally and philosophically indefensible to view *Homo sapiens* as a conqueror species with the attendant right to determine environmental policy solely on the basis of our own human interests. If we agree with this broad conclusion, we might consider thus just what the appropriate role for *Homo sapiens* should be in a non-anthropocentric moral community.

The term that comes almost immediately to mind for such a role is *steward* of the moral community. Much literature has discussed the benefits and drawbacks of this term, but I suggest it here in a way that acknowledges our power to change the environment *and* also requires something significant, and occasionally difficult, of us in return. Such a stewardly role implies a shift from dominion and domination to a managerial, and indeed, ministerial role. A citation from the *Oxford English Dictionary* under the heading of 'Stewardly', mentions this ministerial sense of stewardship thus: "The Government of his Kingdome is not Lordly, but Stewardly and Ministeriall."[11]

[10] Callicott, *Aldo Leopold*, 239.
[11] *Oxford English Dictionary*, (1643 J. COTTON. *Doctr. Ch.2*).

4.1 Problem of the Scope of Intrinsic Value and Stewardship

The ministerial sense of stewardship, rather than dominion, might be argued to follow from the awesome responsibility we have in our demonstrated ability to radically alter the natural environment for our own species-specific purposes. Stewardship can, but does not necessarily, imply anthropocentrism. Instead, it can be fundamentally other-centered, as a position of both privilege and service. Regardless of how we envision the motivation for stewardship, the attendant responsibilities are not significantly in dispute. Further, once we allow that there is a domain of moral community beyond our species boundary, it would appear that this power asymmetry places a burden of responsibility on us to use our power wisely and with equal consideration accorded to whatever we take to comprise the moral community. The argument for environmental stewardship thus is *noblesse oblige*. Sojourner Truth had the point well in hand long ago. "If my cup won't hold but a pint and yours holds a quart, wouldn't you be mean not to let me have my little half-measure?"

4.2 Stewardship and Trust

The above considerations now bring me to the challenge I've been working towards on the question of moral consideration across species boundaries, whether individualistically or holistically conceived. The challenge is how to envision stewardship in a way that acknowledges, and acts appropriately toward, the moral community. This challenge, I believe, faces all who embrace a wide view of the moral community. I would like to propose that the concept of *trust* in stewardly relations gives us a way to discuss humans' moral consideration of, and for, nonhuman entities.

Another citation from the *Oxford English Dictonary* mentions a connection between stewardly behavior and trust in the English language. We read: "If abused that he do not perform his Stewardly trust as hee should, the people…are to look to it."[12] Here we isolate a concept of stewardship that sounds like it is borne in fiduciary terms. If we read it this way, we might ask who the parties are in the fiduciary relationship. Who is the steward? Who is the beneficiary? What are their characteristics? What does such "trust" entail?

It seems to me that a trust relationship has a certain density to it—in terms of behavior and perhaps intention—that is not representative of a wide range of species to species interactions. It would then seem to follow that if the trust relationship cannot be instantiated across a wide range of species boundaries, the role of steward would fall equally in its wake. Our responsibility toward nonhuman entities is based on our ability to have a trust relationship with that entity. That is, the role of steward as a fiduciary for the particular species or even a community of organisms would fall if a trust relationship cannot be shown (although the role of steward for a more narrowly defined moral community could be retained, and it could be argued that a

[12] Ibid., (1642 BRIDGE *Wound. Consc. Cured iv, 26*).

wider range of species is instrumental to the interests of that more narrow moral community with which we have relationship). In other words, if stewardship implies trust, and trust relations require a certain density in behavioral and perhaps intentional terms, then the role of steward is bound, like a fiduciary, only to those entities which are capable of this behavioral and intentional density. Such an outlook does not auger well for species beyond *Homo sapiens*.

This dimension of stewardship seems to bring us back to the anthropocentric strategy of limiting moral worth to human beings and seeing the rest of the natural world instrumentally. The problem is not in the naturalistic account of moral consideration, but in the seeming lack of applicability of "stewardship" terms, a failure of "trust" expectations for nonhuman beings. That is, the terms of a "stewardship" relationship do not seem to extend very far beyond the boundaries of the species *Homo sapiens*.

Let us consider an example. What is crucial in the model of a trust relation, beyond boundaries of human-to-human interactions, can be brought out in Jane Goodall's work on the chimpanzees of Gombe Stream in Tanzania. Goodall spent most of her time since 1960 on the shores of Lake Tanganyika, Tanzania, studying the local chimpanzees. On one of Goodall's websites it tells us that, "At first, the Gombe chimps fled whenever they saw Jane. But she persisted, watching from a distance with binoculars, and gradually the chimps allowed her closer."[13] Indeed, there is a photo on the website of Jane reaching out and touching an infant chimp named Flint. Flint was the son of another chimp named Flo. The caption under the photograph reads: "Flo became so *trusting* [that] she allowed infant Flint to approach and touch Jane."[14]

I would like briefly to consider whether this case of Goodall, Flo, and infant Flint actually counts as a case of cross-species trust interactions. The situation of Goodall, Flint and Flo in the Gombe stream appears to match rather well with this complex trust relationship. In trusting, we turn over access to something we care a great deal about protecting, and we do it typically with respect to things over which we normally have control of such access. Flo allowed Jane direct access to Flint, where she previously guarded against such access.

Annette Baier has written influentially on the concept of trust, and in her work she distinguishes cases of trust based on how the trust might be "let down."[15] She sees that crucial to a robust sense of trust is the reaction of betrayal when a trust is let down. This, we might say is a sense of trust as vulnerability rather than reliability. When we trust someone in the sense of reliability, and the trust is let down, our reactions do not typically rise to the level of feeling betrayal; anger, irritation, disappointment, but not betrayal. Trust, on Baier's first approximation, is "accepted vulnerability to another's possible but not expected ill will (or lack of good will) toward one."[16] Betrayal, rather than disappointment or irritation, in being

[13] Jane Goodall Institute.
[14] Ibid., emphasis added.
[15] Baier.
[16] Ibid., 235.

let down, signals a trust that was based on a significant vulnerability regarding access to something one cares about a great deal.

On this view of trust, there is one element that might be missing in the picture of a trust relationship on Gombe Stream. What might be absent is a sense of betrayal from Flo if Goodall had exploited the trust by suddenly seizing Flint and taking him off to be examined by fellow researchers for several weeks. It does seem that such an act would be a betrayal of trust proper, not just in anthropomorphic terms. Still, it isn't clear to me that Flo would experience it as such. Certainly, if she was able to retrieve Flint she would not "trust" Jane again, but it seems plausible to think that for Flo a sense of being the subject of an act of betrayal might not be forthcoming. Or at least it is unclear that Flo would feel "betrayed" in the way that humans register betrayal.

If the possibility of a *betrayal* of trust is not in place, might we doubt whether this is a real trust relationship after all? It seems clear that it is still a trust relationship, even if there are elements of the "language game" that are absent as it is played in Gombe Stream. To refer to a favored analogy of Wittgenstein, if we make certain changes in a game, it doesn't mean we have another game altogether. We might raise the pitcher's mound and it is still baseball; or we might play chess and handicap an opponent (playing without a Queen, e.g.), but we cannot remove checkmate from the game of chess. Nor can we play tennis without a net. The removal of betrayal in the case of Goodall and the chimps is more like a player in chess being handicapped without a Queen than playing tennis without a net. Flo can play the game of trust with Jane, but like playing chess without a Queen, Flo is always under a handicap to Jane when Flo negotiates access to vulnerabilities. However, the negotiations are still present. Recall that negotiation regarding perceived vulnerabilities between the parties is a crucial element in the implicit scheme of trust.

A major point of these remarks about Goodall and the chimps is to impress upon the reader just how behaviorally (and intentionally?) dense a trust relationship becomes. In addition, it is clear that the relationship will dwindle as the elements of the relationship drop away. It is simply a matter of *which* elements drop away. We can play chess without a Queen, but we cannot really play chess without the concept of checkmate. Some elements of the game—or the relationship—are central and decisive in themselves. Consequently, once we start to move away from higher mammals, the possibility of a trust relationship with an individual of another species begins to vanish in exceedingly rapid terms. I see little prospect of a trust relationship with lizards for example, and much less to insects.

Perhaps some might wish to switch to a more holistic model of intrinsic moral value, as in a land ethic, and avoid all these questions as they apply to individuals. Or perhaps some might wish to reconfigure "trust" such that its terms are appropriate to the capacities of the subjects involved in the relationship. Either way, it would seem to me that the human species, as a citizen of the broader community, ought to view itself as steward of that greater moral domain. Those with great potential to harm must tread carefully and be aware of that power, *especially* if other species and natural entities are understood to have value in themselves. In terms of human action toward the environment, the argument is the same for both holistic and individualistic perspectives: *noblesse oblige.*

5. CONCLUSION

In this essay I have set out to stimulate a way of considering intrinsic moral worth across species boundaries from a naturalistic perspective devoid of "disengaged ontological and metaphysical solutions" and "ideological dogmatism." I suggested two movements: First, it is not clear that our moral consideration should apply only to *Homo sapiens;* second, we are naturally inclined to "care" about others to some degree, and that these impulses are often socially shaped. Still, as Quine notes, "there is no mistaking the grading off of altruistic impulse as we move outward from kin."

The response to this problem has been sketched by introducing an Aristotelian injunction that if we must do something, we do better if we do it well and wisely. The point is that by responding well to our drives to benefit others—and perhaps extending them beyond what might seem "encoded"—we improve our own chances for living well ourselves. Such a model would seem to imply a role for *Homo sapiens* as environmental steward, charged with the ministerial management of the community in which we are environmental citizens. This is, of course, fundamentally rooted in our own self-interests. Indeed, the fiduciary concept of steward seems to bring back a whole host of anthropomorphic (instrumentalist) strategies to assign value to nonhuman entities. This does not mean that stewardship is exclusively anthropocentric (indeed, it can be otherwise); but nonetheless this essay concludes with a challenge to the role of steward insofar as it has fiduciary and trust-based tendrils.

My point in this essay is simply to say that human interests can be seen as linked to the intrinsically valuable interests of nonhuman species or natural entities. The stewardship relation may or may not be helpful in this regard; I have tried to problematize certain assumptions implicit in a stewardship model of relationship. Perhaps the intrinsic value of other species can be recognized and, hopefully, incorporated into arguments about the self-interests of humans in light of the expanding moral community. To enact environmental policy commensurate with this view would be very stewardly indeed.

Acknowledgements: An early version of this paper was presented to Yale University's Interdisciplinary Bioethics project. I wish to thank them for a fruitful discussion; thanks especially to Christiana Peppard, James Fleming, Carol Pollard, and Julius Landwirth for comments and important critical suggestions on this essay.

REFERENCES

Baier, Annette. "Trust and Antitrust." *Ethics* 96 (January 1986): 231-260.

Bentham, Jeremy. *Introduction to the Principles of Morals and Legislation*, chapter 17.

Callicott, J. Baird. In *Aldo Leopold, A Sand County Almanac*. New York: Ballintine Books, 1970.

Jane Goodall Institute. http://www.janegoodall.ca/jane/jane_bio_gombe.html.

Oxford English Dictionary, 2nd Edition. Edited by John Simpson and Edmund Weiner. Oxford: Oxford University Press, 1989.

Plato. *Republic*. A standard collection in which the *Republic* can be found is *The Collected Dialogues of Plato* (edited by E. Hamilton and H. Cairns). Princeton: Princeton University Press. 1961

Quine, W. V. *Quiddities. An Intermittently Philosophical Dictionary*. Cambridge, MA: The Belknap Press of Harvard University Press, 1987.

Singer, Peter. *Animal Liberation*. New York: Avon Books, 1977.

Singer, P. "All Animals Are Equal." In *Morality and Moral Controversies*. Sixth edition. Edited by John Arthur. Upper Saddle River: Prentice Hall, 2002.

GEORGE M. WOODWELL

SCIENCE, CONSERVATION AND GLOBAL SECURITY[1]

A NEW WORLD

We live in a new world in which there is no time or space for war or the threat of war.[2] This new world is beset by a series of global environmental crises that have every sign, if neglected, of being as destructive of civilization and the human future as the nuclear Armageddon we have spent half a century and the world's wealth and time avoiding. There is an immediate need for a relaxation and ultimate elimination of dependence on oil and other fossil fuels for both political and environmental reasons. There is a parallel need for a universal effort to restore the physical, chemical and biotic integrity of the biosphere before biotic and economic impoverishment overwhelm us.

The barbarism of the 11th of September revealed suddenly just how small our world is. And, just as abruptly, we discovered that civilization will prevail only as a result of unified purpose in providing the same security we have enjoyed in the western world to all the nations and among all as individuals. I believe that "peace and security require that all three legs of government function properly: the political system with all its checks and balances, the economic system with the full panoply of regulation it requires, and now, in a world of rapidly intensified demands on all resources, the environmental scientific system."[3] This caveat applies not just for the United States but also for all nations in this now crowded world.

Tony Blair's early October 2001 address to the British Labour Party called for an aggressive response to terrorism but also, more significantly in the long run, he sought to advance a political agenda for the new millennium designed to acknowledge and correct the gross disparities in human welfare around the world: the crisis of Africa, the urgency of economic reform to reverse the polarization of wealth, and the necessity for ratification of the Kyoto Protocol. The lesson was most unfortunately lost on the United States, which cast a cloud over the decennial Earth Summit held in Johannesburg in late August and early September of 2002 by refusing to participate at all if climatic disruption were on the agenda. In doing so the U.S. administration revealed itself as thoroughly committed to prolonging the

[1] Adapted from the acceptance speech for the 2001 Volvo Environmental Prize, October 2001, Gothenberg, Sweden.

[2] Woodwell, "World enough and time?"

[3] Ibid.

A. W. Galston and C. Z. Peppard (eds.), Expanding Horizons in Bioethics, 221-232.

fossil fuel age despite the political and environmental consequences discussed below. Margaret Beckett, Secretary of State for Environment, Food and Rural Affairs in the U.K., added her vivid insights to the ringing challenge from Blair:

> The devastating tragedy which overshadows this conference and all our lives is a sharp reminder of how much we are one world. It reminds us, as Tony [Blair] said recently, that the most basic of human rights is the right to life. But it reminds us too that the existence and the enjoyment of a right to life depend on having the means to sustain that right. It means having air you can breathe, water you can drink, food that's safe to eat.... But the recognition of climate change and its effects has brought much wider understanding of our mutual vulnerability, as people of one planet. It's brought a recognition that we have to take into account the impact of what we do and can do in one part of the globe on what we do and can do in others. The clearest evidence of that wider recognition was displayed in Bonn in July. Ministers from across the world commented with awe on the unprecedented and historic nature of the agreement made... 180 countries signed up to the practical implementation of the climate change programme whose wider principles were agreed at Kyoto. All had to give for any to gain. And we all did... And because we reached agreement in Bonn, we can move on, to push for ratification—and entry into force—of the Kyoto protocol.[4]

It is clear now that the world must move quickly not only to implement the Protocol, as will soon occur without the U.S., but also to move well beyond it to advance renewable energy globally. The topics of Protocol ratification and renewable energy were the subject of vigorous discussions by the large non-governmental community present at Johannesburg. The vigor and strength of the non-governmental meetings made it clear that the tepid governmental discussions, dominated by U.S. apostasy on all environmental matters, were incompetent and largely irrelevant.

In a political climate dominated by bellicosity in response to September 11th, concerns about global ecology and the human-caused disruption of climate and destabilization of environment may seem trivial. But we are in a new world, new not only in its potential for global terrorism, but also new in human potential for good and ill; in the speed and flow of information; in concepts of right and wrong; in concepts of government; in the hopes and expectations of the public; and in the concerns of our political leadership. We have watched the tragic consequences of the hijacking of our airliners for murderous purpose, and we are clear that there are no limits to the antagonism aimed our way. While we cannot allow that event, and the obvious threat of more to come, to pass unchallenged, neither can we allow our response to amplify the vandals' destruction. Our reaction must not hijack the planet into a suicidal plunge into global war. Nor can we allow negligence to produce a suicidal plunge of only slightly longer duration by deflecting or stopping progress against the global threats of climatic disruption and biotic impoverishment.

[4] Beckett.

HOPEFUL STEPS: THE FRAMEWORK CONVENTION ON CLIMATE CHANGE AND THE KYOTO PROTOCOL

In recognition of the threats to human welfare presented by human-induced global climatic disruption, world political leaders at the United Nations Earth Summit meetings in Rio de Janeiro in 1992 signed the U.N.-drafted treaty known as The Framework Convention on Climate Change. That treaty deals with a global problem that is more complicated technically, scientifically and politically than any previous treaty. It says, in sum, that it is the intention of the nations to stabilize the heat-trapping gas content of the atmosphere at levels that will protect human interests and nature. The Treaty was later ratified by more than 180 nations, including the United States, and thereby became global law. The 1997 Protocol to the United Nations Framework Convention on Climate Change was drafted in Kyoto to implement the Treaty. In 2004 the Protocol is still en route to ratification and implementation.[5]

Under the 1997 Protocol, the nations agreed to a complicated formula that would reduce the emissions of the industrialized world about 5% below 1990 levels by the period 2008-2012. The initial reductions were accepted by the industrialized nations, who were the principal polluters. The developing world was not included in the reductions. The United States participated intensively in the preparation of the Kyoto Protocol, which was written substantially to accommodate U.S. interests.

Our sudden withdrawal of support from the Protocol in the first few months of the George W. Bush administration was an astonishing and irresponsible reversal of U.S. policy. While that administration advanced a "program" for nominally addressing the requirements of the Convention, replacing the Protocol, the program actually called for an increase in the use of fossil fuels above current levels, not the reduction below 1990 use agreed to by the U.S. in negotiating the Protocol in Kyoto in 1997.

The United States' withdrawal from the treaty is the greater scandal in that it was the U.S. scientific community that defined the problem of climate change and made it a global public issue. In fact, the effort by the scientific community to put the issue before governments, successful as it has been, had an incubation period reaching back more than three decades to the preparations for the 1972 Stockholm Conference on the Human Environment. At that point, the scientific community acknowledged the problem of the accumulation of heat-trapping gases in the atmosphere; but they saw no measurable change in the temperature of the earth and decided (strangely and over objections from some) that there was no clear basis for recommending action to stop the trend. Instead, the issue was merely discussed and redefined. It took nearly another decade for the issue to gain attention in political circles when, in 1979, the scientific community, with help from the Council on Environmental Quality in the Executive Office of the President, provided sufficient public pressure that Congress held hearings on the threat of climatic disruption.[6]

[5] For the Protocol to become effective without U.S. participation, Russian ratification is necessary. In mid-2004 Russian ratification seems imminent.

[6] At the request of J.G. Speth, then Chairman of the Council on Environmental Quality in the Carter Administration, a statement was prepared calling attention to the seriousness of the threat of global

Despite those hearings, and others that followed, there was no concerted governmental response for another decade. In 1991-1992 the Framework Convention on Climate Change was drafted (at the Earth Summit) in response to a directive by the U.N. General Assembly, which was responding to the concentrated efforts of the scientific community. And it is the scientific community that has held steadfastly ever since to the objective of implementing the treaty.

Steps toward implementation were finally taken in 1997 in Kyoto and formulated in the Kyoto Protocol, described above. While the United States objected to the exclusion of developing countries from emissions standards, experience has shown that the developing world is moving even more effectively than the developed world toward meeting the objectives not only of the Protocol but also the Convention.[7] It is not surprising to discover that solar energy offers shortcuts to economic development in nations such as India and China, and that it can displace oil and coal and the technology associated with fossil fuels under many circumstances.[8] These steps have been taken effectively despite the tortured logic and outright lies of segments of the fossil fuel industry and its allies, and the stunning stupidity of the reversal of the U.S. position in rejecting an agreement painstakingly negotiated among the U.S. and more than 180 other nations.

Yet even the Kyoto Protocol is not enough.[9] The Protocol reflects the negotiations among the nations assembled in Kyoto in 1997. It offers a very small increment toward what is required to meet the details of the Treaty. It bows not at all to what scientists have been saying for more than two decades about the seriousness of the effects of a continued buildup of heat-trapping gases in the atmosphere.[10] It delays until 2008-2012 achievement of a reduction of 5% below 1990 emissions of carbon dioxide. Stabilization of the heat-trapping gas content of the atmosphere, as required under the Framework Convention, would involve a reduction of emissions of 50-60%, and the percentage reduction required is rising annually. The current atmospheric burden of about 379 ppm carbon dioxide (for the year 2004) is beyond levels at which scientists are confident that they can anticipate the effects. Allowing the burden to drift higher only amplifies the problem, raising further questions as to rates of sea level rise, the intensity and locales of climatic disruption, and the implications for climate and virtually all other aspects of the human habitat. The melting of the sea ice in summer in the Arctic Ocean is only one example—which turns the Arctic Ocean into an energy-absorbing black body, instead of a reflective white body under the continuous summer sun. That change is underway and far advanced.

---

climatic disruption. See Woodwell et al., "The Carbon Dioxide Problem." The statement was circulated widely by the Council under Mr. Speth's leadership and became the basis for hearings before the Senate Committee on Public Works.

[7] Ramakrishna and Jacobsen, eds.

[8] See David Goodstein's essay, "Running Out of Gas," for a more detailed discussion of the sustainability of different sources of energy. *Ed.*

[9] See, e.g., Stewart and Wiener.

[10] Woodwell et al., "Biotic Feedbacks."

CLIMATIC INSTABILITY AND ENVIRONMENTAL IMPOVERISHMENT: CORE ISSUES

The consequences for climates of an increased carbon dioxide atmospheric burden are virtually unpredictable, beyond instability. The effects can be as fully devastating as war. The current, continuing, drought in North America and central Asia is consistent with long-standing predictions of the continental effects of the climatic disruption, although proof of cause and effect is never perfect in such matters. Several million people in central China are threatened now with starvation due to the prolonged drought, precisely the type of change in climate and effects anticipated as the earth warms and continental centers dry out.[11]

The instability of climate is but one of several environmental trends, each of which individually has the capacity to disrupt civilization—no less than the threats of war and political chaos that regularly grip the world. The fact is, however, that these trends are underway and the processes are far advanced. Unchecked, they lead inexorably to the biotic impoverishment of the earth,[12] to the economic impoverishment of all, and to political chaos. As destructive environmental effects progress, they quickly multiply the difficulties of maintaining stable and effective governments that are capable of reversing the trends and preserving both human welfare and the opportunity for a working democracy.

The causes of environmental impoverishment are well known:

The Growth of the Human Population

The earth now has a human population of about 6.3 billion. It is increasing annually by about 85 million, which equals an increase of one million people every 4 days. The growth in human numbers places new pressures on land, forests, fisheries, and governments from all sides. These pressures result in political unrest, as increasing numbers of people seek to migrate from poverty to wealth, from tyranny to democracy, from squalor to order. Indeed, never have there been so many migrants across so many borders, including the southern frontier of the United States, the Mediterranean frontier of Europe, and the Pacific borders of China. The United States has been an especially desirable objective, of course, with our stable democratic government and our high standard of living.

But the world is far from helpless in addressing the core issues of population. A very wise major advance was made toward the empowerment of women as a fundamental step toward population control at the Cairo Conference of 1994, which declares that

[11] Intergovernmental Panel on Climate Change, *Climate Change 1995.* See also information provided by the UN Convention to Combat Desertification, www.unccd.int; and The Earth Policy Institute, which has recently issued statements about desertification, available at http://earth-policy.org/Updates/Update23.htm.

[12] See Chapter 4 in Woodwell, *Forests in a Full World.*

> [t]he key to this new approach is empowering women and providing them with more choices through expanded access to education and health services, skill development and employment, and through their full involvement in policy- and decision-making processes at all levels. Indeed, one of the greatest achievements of the Cairo Conference has been the recognition of the need to empower women, both as a highly important end in itself and as a key to improving the quality of life for everyone.[13]

Further advances are possible, if we have the will.

Biotic Impoverishment

We hear about the loss of species, an irreversible change in the global potential for support of life. But long before species are lost, the natural communities that have dominated every corner of the earth have been impoverished to the point of dysfunction. Forests have been reduced to shrublands; shrublands to grassland and persistent herbs; grassland to wasteland or to barren ground. With those changes come dysfunctional landscapes, silted and poisoned rivers, floods and droughts, eroded landscapes and poverty. Parallel impoverishment marks the transitions in aquatic systems from those supporting large-bodied, slowly reproducing plants, fish and mammals to those supporting small-bodied, rapidly reproducing invertebrates and plankton, including toxic forms.[14]

We need not look far for examples. Haiti, the most impoverished nation in the western hemisphere, has less than 3% of its land remaining in forest, row-crop agriculture on 30+ degree slopes, no reliable public water supply, no irrigation despite extensive water works once engineered by the French and others, and abysmal poverty. A government requires a place to stand, resources to work with, not a gridlock of impossible environmental problems. The recent insurgencies and political instabilities in Haiti only emphasize this point. The only solution here is outside help, far beyond the 30% of food supplied through USAID. If we are bold enough and wise, it will require 10–50 billion dollars over a decade or more to implement a plan acceptable to the public for restoration of a landscape that can support people and a government. The landscape must have reliable rivers that flow in defined channels, forested mountain slopes that are stable, fisheries that have recovered from the effects of massive siltation, and a viable agricultural system on the best agricultural land reclaimed from under municipalities and slums.[15]

Without outside help, there is no way that such a transition can proceed in time to aid current generations of people. We cannot allow other nations and the world itself to slip into such disarray, but the process is underway and conspicuous in

[13] United Nations Population Fund.

[14] There is a large and somewhat misleading literature on biodiversity and its importance to human welfare. See for instance, Wilson, *The Diversity of Life*; and Wilson and Peter, eds., *Biodiversity*. The emphasis on biodiversity is misleading because biotic impoverishment precedes the extinction of species and is the cause of environmental breakdown and human impoverishment. The extinction of species follows if the impoverishment is prolonged and widespread.

[15] The perspective on Haiti has been compiled over years from personal experience with the U.S. Department of State, reports of USAID, World Bank and other economic development agencies as well as limited experience with Haitian officials in Haiti.

virtually every nation. There is no help for the world on the moon, or on Mars. We have to help ourselves, and the time is now.

Toxification

Human activities are changing the chemistry of the whole earth. The global carbon, nitrogen and sulfur cycles are intrinsic to all life and are now dominated by human activities.[16] The disruption of the carbon cycle is the basis of the climatic disruption. The massive changes in the nitrogen and sulfur cycles are causes of the pollution and ultimate impoverishment of terrestrial and aquatic plant and animal communities worldwide. In addition to these disruptions of the natural cycles to which all life is adapted, modern industry has produced and released into the biosphere millions of tons of exotic molecules, many of which, such as DDT, have been used because of their biological effects and have now become virtually ubiquitous. Their toxic effects reach far beyond their original purpose and contribute to the poisoning of land and water globally. Indeed, DDT and allied toxins used in agriculture and public health to control vectors of diseases have become an intrinsic part of virtually all life. This is true on a global level. The effects are profound and range from cancer and metabolic and developmental anomalies in individuals to the biotic impoverishment of land and sea.[17]

Climatic Disruption Caused by Global Warming

The most powerful evidence of the failure of the human habitat is the global destabilization of climates by the accumulation of heat in the atmosphere. The facts have not changed fundamentally since 1889, when Svante A. Arrhenius famously recognized that carbon dioxide exerts a warming effect on the global atmosphere. He predicted that doubling the atmospheric levels of carbon dioxide would result in a five to six degree (Celsius) temperature increase globally. The effect of global warming is open-ended in that it will continue until substantial reductions are made in the global use of fossil fuels. The effect is also a positive feedback system: the warming speeds the warming by slowing the absorption of carbon dioxide into the surface water of the oceans and by speeding the release of additional carbon dioxide from organic matter stored in soils and in peat.

But contrary to concerns during Arrhenius' time, the issue now has a political focus: how to achieve the stabilization of the composition of the atmosphere at a

[16] These topics have been treated in various texts in ecology, such as that of Schlesinger, *Biogeochemistry: An Analysis of Global Change*. These issues continue to be the subject of significant research around the world.

[17] The literature documenting the ubiquity of a global contamination is extensive. One of the most persuasive studies of a locally contaminated food web is that defined for eastern Long Island, NY, where spraying with DDT had been done for years to control the salt marsh mosquito. DDT residues appeared at close to acutely lethal concentrations throughout the food web. See Woodwell, Wurster, and Isaacson, "DDT Residues." For a case of the effect of pesticides and other industrial effluents on reproduction, see Steingraber, *Having Faith: An Ecologist's Journey to Motherhood*.

level that will protect human interests and nature as agreed to under the U.N. Framework Convention on Climate Change. That level is probably closer to 300 ppm than to the 379 ppm of 2004. Stabilization at the present concentration would require a reduction in present emissions globally of 3 to 4 billion tons per year of the approximately 8 billion tons currently released from all sources. Such a reduction would entail either a 60% reduction in fossil fuel use immediately or a 50% reduction and a complete cessation of deforestation for agricultural purposes. The reductions would be followed in subsequent years by a need for further reductions over a few decades to a century leading to the ultimate elimination of fossil fuels as a source of energy.[18]

ANALYSIS OF TRENDS AND THE CONCEPT OF THE BIOSPHERE

The question of what effects these trends will entail is much debated. A heavy reliance on models that incorporate physiological responses to the increases in carbon dioxide in the atmosphere produces a most optimistic view of a world with increased accumulation of carbon in lush communities that migrate with climate.[19] Others, including this author, offer a somewhat less stereotyped analysis based on experience and a consideration of a wider range of factors not easily incorporated into models.[20] That analysis shows a series of transitions more akin to the biotic impoverishment discussed above as morbidity and mortality of dominant trees affect forests, insect outbreaks become common, and other pathogens flourish in stressed plants. Data and experience are abundant[21] and more accumulate daily as drought and fires spread across the northern hemisphere continents.

These trends, as seriously threatening as they are, point to one essential transition that might emerge from this most frightening moment in human affairs. It is the recognition that civilization, i.e. the entire advance of the human enterprise globally over the three million years or so of recent human evolution (especially including the most recent 10,000 years of gradually accumulating historical record), has depended on the integrity of function of an environment hospitable to human life, best characterized as "the biosphere." I use the term inclusively to reach to the limits of life on earth, from the stratosphere (which may contain microbial spores, the dust of life), to the limits of life in the depths of the earth. This concept of the biosphere was used by G. Evelyn Hutchinson, by me, and by others in a well-known September 1970 edition of the *Scientific American* devoted to that topic under that title.[22] The most essential feature of the biosphere is that it is a living system maintained by life processes themselves. The reality and importance of that observation is conspicuous now in the accumulating global failure of that system.

[18] Intergovernmental Panel on Climate Change, "The Scientific Basis."

[19] United States Global Change Research Program, National Assessment Synthesis Team.

[20] See the essay in this volume by David Ehrenfeld, entitled "Unethical Contexts for Ethical Questions," which discusses the importance of broadening the context when assessing various new technologies and their effects. *Ed.*

[21] Woodwell and Mackenzie.

[22] Woodwell, "The Energy Cycle of the Biosphere."

The dysfunction of the biosphere results in an environment that is changing quickly and drastically at the very moment that we are reaching out to meet the needs of soaring human numbers and expectations. The consequences of such destruction for global, national, and individual personal security are no less threatening than those of war. Indeed, they are in fact a cause of war as the vise of environmental impoverishment closes. This trend will only intensify as larger swaths of once-fertile land become arid and water supply problems increase.

The global environmental squeeze is the global integration of specific local failures around the world. It is a clear sign that we need to look around ourselves, our lives, our houses, farms and municipalities and nations and re-adjust our activities and use of resources to conform to a set of standards that, when summed to the world as a whole, re-establishes a stable and sustainable biosphere. We must widen the context beyond our immediate neighborhoods. Re-establishing the dominance of natural ecosystems in management of the earth is a major task; but it is the only path that can work. Preserving the earth as a self-maintaining, regenerative living system is the emergent, essential objective. It is more important than war, for, failing, there is nothing worth fighting over. It is important enough to be a basis for challenging not only human activities but also inventions, technologies, and even dreams.

INTEGRITY OF EARTH AND HUMAN ACTION

There is ample precedent for such worldwide imperatives in law and in human affairs. It is unacceptable for example to murder one's fellow citizens by spreading mercury over the landscape, or to make children stupid by exposing them to lead, or to distribute DDT in the United States. It is a small step to move from protecting personal security from poisons distributed by one's neighbor, or by industry, to protecting the security of all by managing landscapes and regions to preserve their *physical, chemical and biotic integrity*. In fact, these very words have been a part of the objectives of every incarnation of the Water Pollution Control Act in the United States since 1972: *physical, chemical and biotic integrity.* It is difficult to exaggerate the importance of such "integrity."

The key elements to maintaining this integrity are energy and forests, and both demand attention now. In terms of energy, there is a clear connection to the present world crisis as the United States moves to protect its interests in access to Arabian oil and attempts to increase its own domestic production despite a lack of reserves. But the infatuation of the industrialized world with oil is a cause not only of climatic disruption, political instability, and bellicose bluster punctuated by occasional outbreaks of war but it is also the cause of a host of serious pollution problems such as the acidification of rain with oxides of sulfur and nitrogen. The threat to security is double-edged. There is an immediate economic threat if oil is cut off, and a slightly less immediate—but real and global and fatal—environmental threat if it is not. What is required is an awareness of long-term possibilities, not just short-term market incentives.

Forests are the second component to biospheric integrity. They are so large in the world in area, in carbon content, and in influence on global and local energy and water budgets, that we must think of them as the great biotic flywheel that keeps the biosphere functioning as a stable human habitat. Deforestation, the change in land use from forest to agriculture or, ultimately, barren land, contributes 1.6 to 2.0 billion tons of carbon annually to the total of about 8 billion tons released annually through human activities.

If we are to reduce carbon emission levels as much as the 3-4 billion tons mentioned above, we must pay attention to forests as well as fossil fuels, if only because forests are the natural vegetation of such a large fraction of the land area. Forests originally covered about 44% of all land. They have been reduced to about 28% now but still exert a very large influence on energy, water, and climate regionally and globally. The absolute protection of the earth's remaining primary forests, most of which lie in the tropics of Africa, in the Amazon Basin, Borneo and Siberia, is essential in moving toward controlling and stabilizing the composition of the atmosphere. Attention to forests is a feasible goal: restoration of deforested lands is a step in restoring the functional integrity of landscapes such as Haiti, as well as other impoverished and eroding drainage basins around the world. The efforts begin at home, but they ultimately sum to a biosphere that is either functional and has a future, or is progressively dysfunctional and a certain cause of continued political instability and spreading human misery.

While the world will see many causes of the immediate crises of environment and government and also many solutions, the ecologists are not mistaken in their recognition of a chain of dependencies between human welfare and the fundamental resources of air, water, land and a place to live. We expect our governments, at least in democratic societies, to establish and defend equity in access to those essential resources. Indeed, it is a core function of government. While the urge to stamp out terrorism is correct, there is always going to be desperate resentment in a world in which there continues to be an increasing and soaring differentiation of rich and poor, of haves and have nots, of equity and lack of equity in opportunity to live in safety and comfort under well-regulated laws. Despite the necessity for a major global effort in controlling and (if we are persistent and fortunate) eliminating terrorism, nothing has changed the urgency of addressing the decay of the human environment through climatic disruption and biotic impoverishment.

Our concern is long standing and consistent: climatic disruption through human-caused changes in the composition of the atmosphere will only provoke further troubles in the world. Integrity is needed in human action in order to restore integrity to the environment in which we all live and upon which we all depend. The environmental basis of this concern is rooted in science that has a century and more of research behind it. And there are, despite persistent critics, abundant recent data confirming the transitions and plentiful new insights into the global bioclimatic system.

Time is short. The world is already at levels of heat-trapping gases that will produce effects outside the realm of predictability and therefore outside the realm of acceptability or reasonable risk. The global transition of the 11th of September only makes the issue more urgent, not as some would have it, less.

CONCLUSION

We come to the objective: a massive shift away from fossil fuels, toward locally available renewable sources of energy, and toward the restoration of the functional integrity of land and water as essential to continued human habitation of the biosphere. Both are essential to human security and to the independence, self-sufficiency and security of individuals and nations.

The transition need not be immediate; it cannot be. It requires public leadership and, ultimately, governmental responsibility and support. But the opportunity to make that transition is here. It can start with a young and vigorous scientific community, just as the Framework Convention started with some scientific revolutionaries who held meetings around the world, and who ultimately persuaded the United Nations General Assembly to proceed with drafting a treaty which is now the law of the world.

Further, we need a new set of innovations to bring an immediate 20% reduction in use of fossil fuels nationally by the United States and other industrialized nations, and to advance the restoration of the functional integrity of the biosphere. The local, national and global responsibility of this generation is essential to ridding the world of all forms of terrorism, degradation and destruction.

All have been left reeling by recent violent events and the continued march of uncertain military and political sequelae. There is an overwhelming sense that we have experienced a major transition in the globalization of the human endeavor, and the urgency of the environmental transition has become only more pronounced. The decade of experience with the Framework Convention on Climate Change and the Kyoto Protocol, followed by the Johannesburg Summit, started with the momentum of one of the most promising treaties of all time. Yet it ended with a scandalous rejection of the Protocol by a new U.S. administration committed to the oil industry. While the official meetings in Johannesburg became virtually irrelevant under U.S. leadership, it became clear that the Kyoto Protocol will enter into force even without the U.S., and that the rest of the world is acutely aware of the emergence of the multiple crises of environmental degradation. A vigorous non-governmental community at Johannesburg captured the essence of the moment as a challenge to the scientific and political communities to advance a genuine revolution in the human undertaking, replacing *a strategy of failure* based on the corruption and impoverishment of the human habitat with a *strategy of hope* based on the biosphere's potentially infinite capacity for renewal and self-repair. The place to start is with the implementation of the intent and details of the Framework Convention on Climate Change, already the law of the world.

REFERENCES

Beckett, Margaret. Speech given at Labour Party conference, Brighton, U.K., October 2001. http://www.defra.gov.uk/corporate/ministers/speeches/mb021001.htm.

Intergovernmental Panel on Climate Change. *Climate Change 1995: Impacts, Adaptations and Mitigation of Climate Change: Scientific-Technical Analyses*. New York: Cambridge University Press, 1996.

________. "The Scientific Basis. Contribution of Working Group I to the Third Assessment of the Intergovernmental Panel on Climate Change." *Climate Change 2001.* New York: Cambridge University Press, 2001.

Ramakrishna, K., and L. Jacobsen, eds. *Actions Versus Words: Implementation of the UNFCCC by Select Developing Countries—Argentina, Brazil, China, India, Korea, Senegal, South Africa*. Woods Hole, MA: Woods Hole Research Center, 2003.

Schlesinger, W.H. *Biogeochemistry: An Analysis of Global Change*. Durham: Duke University Press, 1997.

Steingraber, Sandra. *Having Faith: An Ecologist's Journey to Motherhood.* Cambridge: Perseus Books, 2001.

Stewart, R.B. and J.B. Wiener. "Practical Climate Change Policy." *Issues in Science and Technology* (Winter 2004): 71-78.

United Nations Population Fund. "Summary of International Conference on Population and Development." http://www.unfpa.org/icpd/icpd.htm.

United States Global Change Research Program, National Assessment Synthesis Team. *Climate Change Impacts on the United States: The Potential Consequences of Climate Variability and Change.* New York: Cambridge University Press, 2001.

Wilson, E.O. *The Diversity of Life.* Cambridge: The Belknap Press of Harvard University, 1992.

Wilson, E.O., and F.M. Peter, eds. *Biodiversity*. Washington, D.C.: National Academy Press, 1988.

Woodwell, G.M. "The energy cycle of the biosphere." *Scientific American* 223, no. 3 (1970): 64-74.

________. *Forests in a Full World.* New Haven: Yale University Press, 2001.

________. "World enough and time?" *Conservation Biology* 17, no. 2 (April 2003): 356-7.

Woodwell, G.M., G.J. MacDonald, R. Revelle, and C.D. Keeling. "The carbon dioxide problem: Implications for policy in the management of energy and other resources." *Report to the Council on Environmental Quality.* Bulletin of the Atomic Scientists 5, no. 8 (1979): 56-57. Washington, D.C.: Congressional Record.

Woodwell, G.M., and F.T. Mackenzie. *Biotic Feedbacks in the Global Climatic System.* Oxford: Oxford University Press, 1995.

Woodwell G.M., F.T. MacKenzie, R.A. Houghton, M. Apps, E. Gorham and E. Davidson. "Biotic Feedbacks in the Warming of the Earth." *Climatic Change* 40 (1998): 495-518.

Woodwell, G.M., C.F. Wurster, and P.A. Isaacson. "DDT residues in an East Coast estuary: A case of biological concentration of a persistent insecticide." *Science* 156 (1967): 821-824.

DAVID GOODSTEIN

ENERGY, TECHNOLOGY AND CLIMATE

Running Out of Gas

Let me begin by stating my conclusions. We (the world) will soon start to run out of conventional, cheap oil. If we manage somehow to overcome that shock, and life goes on more or less as it has been, then we will start to run out of all fossil fuels before the end of this century. In that case, there is a very real chance that by the time we have burned up all the fuel, we will have rendered the planet uninhabitable for human life. And even if human life does go on, civilization as we know it will not survive, unless we can find a way to live without fossil fuels.[1]

Technically, it should be possible to accomplish that. Stationary power can be obtained from nuclear energy and from light from the sun. Part of that power can be used to generate hydrogen fuel for use in transportation. There are technical problems to be solved, certainly, but the scientific principles are all well understood, and we are very good at solving technical problems. In fact if we put our minds to it, we could kick the fossil fuel habit now, protecting the planet's climate from further damage, and preserving the fuels for future generations to use as the source of chemical goods. To do that would require political leadership that is both visionary and courageous. It seems unlikely that we will be so lucky.

Thus, we are faced with a grave crisis that may change our way of life forever. We live in a civilization that evolved on the promise of an endless supply of cheap oil. The era of cheap oil will end, probably much sooner than most people realize. To put this looming crisis in perspective, and to judge its significance, it helps to start from the beginning. Here is how it all works.

Nuclear reactions inside the Sun heat its surface white hot. From that hot surface, energy in the form of light, both visible and, to our eyes, invisible, radiates uniformly away in all directions. Ninety-three million miles away, the tiny globe called Earth intercepts a minute fraction of that solar radiation. About 30% of the radiation that falls on the Earth is reflected directly back out into space. That's what one sees in a picture of the Earth taken, say, from the moon. The rest of the radiant energy is absorbed by the Earth.

A body such as the Earth that has radiant energy falling on it warms up or cools down until it is sending energy away at the same rate it receives it. Only then is it in a kind of equilibrium, neither warming nor cooling. In any given epoch, the Earth, like the Moon or any other heavenly body, is in steady state balance with the Sun,

[1] A more complete discussion of the issues raised in this essay can be found in the recent book, *Out of Gas*, by David Goodstein, W.W. Norton (2004).

A. W. Galston and C. Z. Peppard (eds.), Expanding Horizons in Bioethics, 233-245.

neither gaining nor losing energy. That is the primary fact governing the temperature at the surface of our planet.

The rate at which the Earth radiates energy into space depends on its temperature. Because it receives only a tiny fraction of the Sun's energy, it radiates much less energy than the Sun does. So, it can balance its energy books at a temperature much cooler than the Sun. In fact it can radiate as much energy as it receives with an average surface temperature of zero degrees Fahrenheit. The Earth's radiation is not visible to our eyes. It is called infrared, or "below-red," radiation because its color is beyond the red end of what we are capable of seeing.

Fortunately for us, that is not the whole story. If the average surface temperature of the Earth were really zero degrees Fahrenheit, we probably would not be here. The Earth has a gaseous atmosphere. The atmosphere is largely transparent to the white-hot radiation from the Sun. The nitrogen and oxygen that make up nearly all of the Earth's atmosphere are transparent as well to the infrared radiation from the Earth; but there are trace gases, including water vapor, methane and carbon dioxide that absorb infrared radiation. Thus the blanket of atmosphere traps about 88% of the heat the Earth is trying to radiate away. The books remain balanced, with the atmosphere radiating back into space the same amount of energy the Earth receives, but it also radiates energy back to the Earth's surface, warming it to a comfortable average temperature of fifty-seven degrees Fahrenheit. That is what's known as the greenhouse effect.

There is a tiny but vital exception to the perfect energy balance of the Earth-Sun system. Of the light that falls on the Earth, an almost imperceptible fraction gets used up nourishing life. Through photosynthesis, plants make use of the Sun's rays to grow. Animals that eventually die eat some of the plants. Natural geological processes bury some of that organic matter deep in the Earth.

For hundreds of millions of years, animal, vegetable and mineral matter has drifted downward through the waters to settle on the floor of the sea. In a few privileged places on Earth, strata of porous rock were formed that were particularly rich in organic inclusions. With time, these strata were buried deep beneath the sea floor. The interior of the Earth is hot, heated by the decay of natural radioactive elements. If the porous source rock sank just deep enough, it reached the proper temperature for the organic matter to be transformed into oil. Then the weight of the rock above it could squeeze the oil out of the source rock like water out of a sponge, into layers above and below, where it could be trapped. Over vast stretches of time, in various parts of the globe, the seas retreated, leaving some of those deposits beneath the surface of the land. Other theories of how oil originated have been proposed from time to time, but they have not stood up. Modern instruments are even able to detect what sorts of organisms went into making different deposits of oil. Nearly all geologists today agree with this account of how oil came to be.

Oil consists of long molecules of carbon and hydrogen. If the source rock sank too deep, the excessive heat at greater depths broke the hydrogen and carbon molecules into the smaller molecules that we call natural gas. Meanwhile, in certain swampy places on land, the decay of dead plant matter created peat bogs. Over the course of the eons, buried under sediments and heated by the Earth's interior, the peat was transformed into coal, a substance that consists mostly of elemental carbon. Coal, oil and natural gas are the primary fossil fuels. They are energy from the Sun, stored within the Earth.

Not all of the energy from the Sun flows back out into space. A tiny fraction of distilled sunlight gets stored up in the form of fossil fuels. The process is agonizingly slow and inefficient. But it has been going on for an extremely long time. The net result is that the Earth has accumulated a legacy of oil that we in our generation have inherited, discovered, and put to use.

Until only 200 years ago (the blink of an eye on the scale of history), the human race was able to live almost entirely on light as it arrived from the Sun. The Sun nourished plants that provided food and warmth for us and for our animals. It illuminated the day and, in most places, left the night sky sparkling with stars, to comfort us in our repose. A few people traveled widely, even sailing across the oceans, but most people probably never got very far from the villages where they were born. In Europe, the lives of the wealthy were garnished with beautiful paintings, sophisticated orchestral music, elegant fabrics and gleaming porcelain from China. For the majority of people in Europe and around the world, there were more homespun versions of art and music, textiles and pottery. Mercantile sailing ships ventured to sea carrying exotic and expensive cargoes including spices, slaves and, in summer, ice. No more than a few hundred million people populated the planet. A bit of coal was burned here and there for one purpose or another, but, by and large, the Earth's legacy of fossil fuels was left untouched.

The situation has changed dramatically. We now expect illumination at night and air conditioning in summer. We may commute to work every day, traveling up to one hundred miles each way between our homes and offices, and we rely on multi-ton individual vehicles to transport us back and forth on demand. Thousands of airline flights per day can take us to virtually any destination on Earth in a matter of hours. When we arrive at our destination, we can still chat with our friends and family back home, or conduct business as if we had never left the office. In industrialized parts of the world, global commerce ensures that the amenities that were once the purview of the rich are now available to most people. Refrigeration rather than spices preserves food, and machines do much of the work that was once done by slaves, indentured servants, or serfs. Ships, planes, trains and trucks transport goods of every description all around the world. The population of the Earth is approaching ten billion people. We don't see the stars so clearly anymore, but on most counts, few of us would choose to return to the world as it existed several centuries ago.

This revolutionary change in our standard of living did not come about by design. If you asked an eighteenth century sage like Benjamin Franklin what the world really needed, he would not necessarily have described the situations and amenities of which our modern world is composed—except perhaps for the dramatic improvement in public health that has also occurred. Our current world is the result of a series of inventions and discoveries that altered our expectations—not an airtight system of societal design. Our current world is less the result of a long-term vision of modern society than it is the result of what nature and human ingenuity made possible for us.

One consequence of those inventions and changed expectations is that we no longer live only, or even primarily, on direct light from the Sun. Instead we consume the fossil fuels made from sunlight that the Earth stored up over those many hundreds of millions of years. In so doing we have unintentionally created a trap for ourselves. We will, so to speak, run out of gas. There is no question about that, since the Earth's fossil fuel reserves are limited. The question is: When will it happen?

The answer is not simple, but some of those who know best, certain petroleum geologists, predict that the first great crisis will come in this decade.[2] Throughout the twentieth century, demand and supply of oil grew rapidly. These two are essentially equal: oil is always used as fast as it is pumped out of the ground. Until the 1950s, many oil geologists asserted the mathematically impossible expectation that the same rate of increase could continue forever. All cautionary warnings of finite supplies were maligned because new reserves were being discovered faster than consumption was rising. Then, around 1956, a clever and insightful geophysicist named M. King Hubbert predicted that the rate at which oil could be extracted from the lower 48 United States would peak around 1970 and decline rapidly after that. When he turned out to be exactly right, other oil geologists started paying serious attention.

Hubbert used a number of methods in his calculations. The first was similar to ideas that had been used by population biologists for well over a century. When a new population (of humans or any other species) starts growing in an area that has abundant resources, the growth is initially exponential. That means that the rate of growth increases by the same amount each year, like compound interest in a bank account. This logic matched that of the 1950s oil geologists: oil discovery would continue to grow unfettered. However, population biologists also observed that once the population is big enough that the resources no longer seem unlimited, the rate of growth starts slowing down. The same happens with oil discovery: the chances of finding new oil decrease when there is less new oil to find. Hubbert showed that, once the rate of increase of known oil supplies starts to decline, it is possible to extrapolate the declining rate to identify when growth will stop altogether. At that point all the oil in the ground will have been discovered. Further, the total amount of oil in the world is equal to the amount that has already been used, plus the known reserves still in the ground. Hubbert noticed that the trend of declining annual rate of oil discovery was established for the lower 48 states by the 1950s. Others have now pointed out that the rate of discovery worldwide has been declining for decades. The total quantity of conventional oil that the Earth stored up for us is estimated by this method to have been about two trillion barrels.[3]

Hubbert's second method required assuming that in the long run, a graph of the historic record of the rate that oil was pumped out of the ground would be a bell-shaped curve. That is, it would first rise (as it has done); then reach a peak; then decline at the same rate at which it rose. Half a century after Hubbert made that assumption, he has been vindicated in the case of the lower 48. If this theorem is

[2] See, e.g., Hubbert; Youngquist; Campbell and Laherrère; Duncan; Ivanhoe; and Deffeys.
[3] Deffeyes.

correct for the rest of the world, and if we already have the historical record of the positive slope of the curve plus a good estimate of the total amount of oil that ever was (two trillion barrels, as above), then it is not difficult to predict when "Hubbert's peak" will occur. Hubbert had that information in the 1950s for the lower 48 states, and we now have that information for the whole world. Different geologists, using different data and methods, yield slightly different results; but some (not all) have concluded that the peak will happen at some point in this decade. The point can be seen without any fancy mathematics at all. Of the two trillion barrels of oil we started with, nearly half have already been consumed. The peak occurs when we reach the halfway point. That, they say, can't be more than a few years off.

Hubbert's third method applied the observation that the total amount of oil extracted to date paralleled oil discovery (total already extracted plus known reserves) but lagged behind by a few decades. In other words, we pump oil out of the ground at about the same rate that we discover it, but we pump and consume it a few decades after the initial discovery. Thus the rate of discovery predicts the rate of extraction. Worldwide, remember, the rate of discovery started declining decades ago. In other words, Hubbert's peak for oil discovery has already occurred. That gives an independent prediction of when Hubbert's peak for oil consumption will occur. It will occur, according to that method as well, some time in the next decade.

Not all geologists agree with this assessment. Many prefer to take the total amount of oil known to be in the ground, divide that by the rate at which it is consumed, and conclude that we can go on like this for a long time. In the oil industry, this is known as the reserves-to-production (R/P) ratio. Depending on what data one uses, that number is currently between forty and 100 years.

Another point of disagreement concerns the total amount of oil in the world. Over the five-year period leading up to 2000, the highly respected United States Geological Survey (USGS) made an exhaustive study of worldwide oil supplies. The resulting report concludes that, with 95% certainty, there were at least two trillion barrels before any oil was extracted from the earth. The report also concludes there is a fifty-fifty chance that there were at least 2.7 trillion barrels, a number that would leave much more in the ground today. However, that number is based on the assumption, contrary to the trends we discussed earlier, that new discovery will continue at a brisk rate for at least thirty more years.[4] The additional 0.7 trillion barrels would amount to discovering all the oil in the Middle East all over again.

The fact is that the amount of known reserves is a very soft number. For one reason, it is usually a compilation of government and commercial figures from countries around the world, and those reported figures are at least sometimes slanted by political and economic considerations. Also, what we mean by "conventional" or "cheap" oil changes with time. As technology advances, the amount of reserves that can be economically tapped in known fields increases. The way the oil industry uses the term, the increase in recoverable oil counts as "discovery," and it accounts for much of the new discovery the USGS expects in the next thirty years. Finally, as oil starts to become scarce and the price per barrel increases, the amount recoverable

[4] U.S. Geological Survey, World Petroleum Assessment 2000.

at that price will necessarily also increase.[5] These are all tendencies that might help to push Hubbert's peak further into the future than the most pessimistic predictions.[6]

Nevertheless, all of our experience with the consumption of natural resources suggests that the rate at which we use them up starts at zero, rises to a peak that will never be exceeded, and then declines back to zero as the supply becomes exhausted. There have been many instances of that behavior: coal mining in Pennsylvania, copper in northern Michigan, and many others (including oil in the lower 48).[7] That picture forms the fundamental basis of the views of Hubbert and his followers, but it is ignored entirely by those who depend on the R/P ratio. Given that worldwide demand for oil will continue to increase (as it has for well over a century), Hubbert's followers expect the crisis to occur when the supply peak is reached rather than when the last available drop is pumped. In other words, we will be in trouble when we've used up half the oil that existed, not all of it. If you believe the Hubbert's peak theory (that the crisis comes when we reach the production peak rather than the last drop), but accept the USGS estimate that there may have been 2.7 trillion barrels of oil, then, compared to the earlier estimates, the crisis will be delayed by about a decade.[8]

If Hubbert's followers are correct, we will be in for some difficult times in the near future. In an orderly, rational world, it might be possible for the gradually increasing gap between supply and demand for oil to be filled by a substitute. But anyone who remembers the oil crisis of 1973 knows that we do not live in such a world, especially when it comes to an irreversible shortage of oil. It is impossible to predict exactly what will happen, but we can, all too easily, envision a civilization paralyzed and decaying from lack of oil, the landscape littered with the rusting hulks of useless SUVs. Worse, desperate attempts of one country or region to maintain its standard of living at the expense of others could lead to yet another dubious war over oil. Knowledge of science is not useful in predicting whether such dire political events will occur; but science is useful for predicting the limits of natural supply and conceiving of (or ruling out) various fuel alternatives.

To begin with, conventional oil is not the only fossil fuel. Once all the cheap oil is pumped, advanced methods can still squeeze a little more oil out of almost any field. There is also what is known as oil sands or tar sands and heavy oil (essentially, the remains of depleted oil fields). These are deposits of oil that are more difficult and expensive to extract than conventional oil. Next there is shale oil. As we have seen, conventional oil came about when source rock, loaded with organic matter, sank just deep enough in the Earth to be cooked properly into oil. Oil shale, from which shale oil can be extracted, is source rock that never sank deep enough to make oil. There are very large quantities of it in the ground, and it can be mined, crushed and heated to produce an oil-like substance. Another possible fossil

[5] Lynch.

[6] U.S. Department of Energy.

[7] For further examples see Youngquist.

[8] I have made this calculation using Hubbert techniques. Hubbert represented the rise and fall of oil discovery and extraction by a mathematical form known as the Logistic Curve (also known in business schools as the S-Shaped Curve). Others have used different bell-shaped curves known as Gaussian and Lorentzian Curves. They all give approximately the same results.

fuel is called methane hydrate, which consists of methane molecules trapped in a kind of cage of water molecules. It is a solid that looks like ice, but that burns when ignited. Nobody knows how to mine methane hydrate or how much of it there is, but there may be quite a lot of it in deep, cold regions of the ocean.

Exploiting any of those resources will be more expensive and slower than pumping conventional oil.[9] Once past Hubbert's peak, as the gap between rising demand and falling supply grows, the rising price of oil will make those alternative fuels economically competitive, but it may not prove possible to get them into production fast enough to fill the growing gap. That double bind is called the "rate of conversion" problem. Worse, the economic damage done by rapidly rising oil prices may undermine our ability to mount the huge industrial effort needed to get the new fuels into action.

Natural gas, which comes from overcooked source rock, is another short-term alternative. Natural gas (primarily methane) is relatively easy to extract quickly, and transformation to a natural gas economy could probably be accomplished more easily than is the case for other alternative fuels. Ordinary engines similar to the ones used in our cars can run on compressed natural gas. Even so, replacing the existing vehicles and gasoline distribution system fast enough to make up for the missing oil will be difficult. And even if this transformation is accomplished, it is only temporary. Hubbert's peak for natural gas is estimated to occur only a couple of decades after the one for oil.

Alternatively, there is a huge amount of chemical potential energy stored in the Earth in the form of elemental carbon—that is to say, in the form of coal. With coal, as with the other fossil fuels, to extract the stored energy, each atom of carbon must be converted to a molecule of carbon dioxide. Unfortunately, carbon dioxide is a greenhouse gas, and converting mass quantities of coal in the ground into carbon dioxide would have consequences for the Earth's climate that are not entirely predictable.[10] In addition, coal is a very dirty fuel: it often comes with unpleasant impurities such as sulfur, mercury or arsenic that can be extracted from the coal only at considerable expense.

Nevertheless, coal can be liquefied and used as a substitute for oil. If we take our chances on fouling the atmosphere and turn to coal as our primary fuel, we are told that there is enough of it in the ground to last for hundreds of years. That estimate however is like the R/P ratio for oil. It does not take into account the rising world population, or the fact that the rest of the world would like to consume more fuel. Moreover, we now use twice as much energy from oil as from coal, and since the conversion process is inefficient, we would have to mine coal many times faster than we are doing now. Finally, that estimate does not take into account the Hubbert peak effect, which is just as valid for coal as it is for oil. The simple fact is, the end of the age of fossil fuel, coal included, will probably come in this century.

[9] It will also require more energy input to get a given amount of energy out. Once the energy needed gets to be equal to the energy produced, the game is lost. We already use one fuel that requires more energy than it provides: ethanol made from corn is a net energy loser. We use it for purely political reasons.

[10] It would not deoxygenate the atmosphere, however. Burning all the known coal in the ground would consume less than 1% of the oxygen in the atmosphere.

Perhaps we should examine different sources of energy. Controlled nuclear fusion has long been seen as the ultimate energy source of the future. The technical problems that have prevented successful use of nuclear fusion up to now may someday be solved, though probably not in time to rescue us from the slide down the other side of Hubbert's peak. Then the fuel could be deuterium, a form of hydrogen found naturally in seawater, and lithium, a light element found in many common minerals.[11] There would be enough of both to last for a very long time. However, the conquest and practical use of nuclear fusion has proved to be very difficult. It has been said of both nuclear fusion and shale oil that they are the energy sources of the future, and always will be.

Nuclear fission, on the other hand, is a well-established technology. While the very word "nuclear" strikes fear into the hearts of many people,[12] it could be a potentially valuable source of energy. When the oil crisis occurs, the fear of nuclear energy is likely to recede before the compelling need for it. However, once we bite the bullet and decide to go nuclear again, it will take at least a decade before new plants start coming on line. And even then there will continue to be legitimate concerns about safety and nuclear waste disposal. Also, nuclear energy is suitable only for stationary power plants or very large, heavy moving things (ships, submarines). Don't look for nuclear cars or airplanes any time soon.[13]

Economists seem to believe that the energy supply problem is not real. As oil becomes scarce, they argue that its price will rise; this will promote competition and innovation in fuel technologies, permitting other sources of energy to emerge. However, as we have seen, that argument ignores the fundamental reality that fuel technologies take time to develop. Furthermore, our vehicles, our roads, our cities, our power plants, our entire social organization have evolved on the promise of an endless supply of cheap oil; even if a viable alternative were to be discovered and available tomorrow, it seems unlikely that the era of cheap oil will end painlessly.

It is more likely that when the peak occurs, the confluence of rapidly increasing demand and rapidly decreasing supply will be disastrous. We had a small foretaste of what might happen in 1973, when some Middle Eastern nations took advantage of the declining U.S. supplies and created a temporary, artificial shortage of oil. The immediate result was long lines at the gas stations, accompanied by panic and despair for the future of our way of life. But after Hubbert's peak, the shortage will not be artificial and it will not be temporary. It will be permanent. At the very least, the end of cheap oil will mean steep inflation, due in part to the rising cost of gasoline at the pump, but also due to the rising cost of transportation and

[11] The nuclear reaction envisioned for fusion reactors is the fusion of deuterium and tritium, two isotopes of hydrogen. Tritium doesn't exist in nature, but the fusion reaction yields neutrons, which would be used to make the tritium in a lithium blanket. Thus the actual fuels are deuterium and lithium.

[12] So much so that the utterly innocent technique called Nuclear Magnetic Resonance by scientists had to be renamed Magnetic Resonance Imaging before it could be accepted by the public for medical use.

[13] The fuel for this kind of reactor is the isotope uranium 235. Natural uranium consists of about 0.7% isotope 235, and 99.3% isotope 238. The known reserves could produce enough energy to replace all fossil fuels for no more than one or two decades. However, if it is used in a type of reactor called a breeder reactor (that converts the otherwise useless isotope uranium 238 into plutonium 239, which is a nuclear fuel), the supply becomes much larger.

petrochemicals. In fact, 90% of the organic chemicals we use, including pharmaceuticals, agricultural chemicals and commodities such as plastics are made from petroleum. There are better uses for the stuff than burning it up.

Once Hubbert's peak is reached and oil supplies start to decline, how fast will the gap grow between supply and demand? That is a crucial question, yet it is almost impossible to answer with confidence. I postulate the following. The upward trend at which the demand for oil has been growing amounts to an increase of a few percent per year. On the other side of the peak, we can guess that the available supply will decline at about the same rate, while the demand continues to grow at that rate. The gap, then, would increase at about, say, 5% per year. That means that, ten years after the peak, we would need a substitute for nearly half the oil we use today—something approaching 10-15 billion barrels per year. Even in the absence of any major disruptions caused by the oil shortages after the peak, it is very difficult to see how that can possibly be accomplished.

What about the possibility that a huge new discovery of conventional oil will put off the problem for the foreseeable future? Better to believe in the tooth fairy. Oil geologists have literally gone to the ends of the Earth searching for oil. There probably isn't enough unexplored territory on Earth to contain a spectacular unknown oil field.[14] Remember that despite intense worldwide effort, the rate of oil discovery started declining decades ago, and it has been declining ever since. That is why the USGS assumption of thirty more years of rapid discovery, mentioned earlier, seems questionable even if it is more a prediction about future technology than future discovery. But let us suppose for one euphoric moment that one more really big field is still out there waiting to be discovered. The largest oil field ever discovered is the Ghawar field in Saudi Arabia, whose 87 billion barrels were discovered in 1948. If someone were to stumble onto another 90 billion barrel field tomorrow, Hubbert's peak would be delayed by a year or two, well within the uncertainty of our present estimates of when it will occur. In other words, it would hardly make any difference at all.

That fact points up the sterility of our current national debate about the Arctic National Wildlife Refuge (ANWR) in Alaska. If the ANWR were opened for drilling (and if it really contains oil, not water as some geologists suspect), it could yield enough oil to supply the United States for about three months. The best reason for not drilling there is twofold: first, to preserve the oil for future generations to use in petrochemicals, rather than burning it up in our SUVs; and second, to protect the wildlife.

Besides, burning all the fossil fuel in the ground poses another grave danger for us. Every carbon atom in the fossil fuels we burn turns into a molecule of carbon dioxide gas in the atmosphere. Recall that carbon dioxide is a greenhouse gas. We have been pouring it into the atmosphere since we started burning fossil fuels in large amounts during the nineteenth century. The net result of tinkering in that way

[14] The largest remaining area that is accessible and unexplored is the South China Sea. Geologists consider it a promising, but not spectacular region. It is unexplored because of conflicting ownership claims by various nations, and murky international law governing such mineral rights at sea. Other possibilities that come with big problems include central Siberia and the very deep oceans.

with the atmosphere is not easy to predict. Increasing the amount of carbon dioxide increases the amount of infrared radiation intercepted by the atmosphere and radiated back to Earth. That warms the Earth slightly, causing more water to evaporate. Water vapor is a powerful greenhouse gas, so the effect of the carbon dioxide is amplified. The warming also causes the polar ice caps to shrink, reducing the amount of sunlight reflected directly back to space, which leads to even further warming. On the other hand, the extra moisture in the air tends to condense into more clouds, and clouds reflect sunlight, decreasing the warming effect. And so on. The effect is complex, with both positive and negative reinforcement acting in ways that will have consequences that are not well understood, though theories abound. There is even a theory that the melting of the polar ice caps could lead to a change in the ocean currents that would cause a sudden cooling of the entire planet.

The cozy climate we now enjoy is in what scientists call a metastable state. That means small perturbations do not cause drastic changes. But it also means we could tumble out of that state into a completely different one. There are other possible metastable states that are dramatically different from the one we're in now, without changing the central feature of the Earth's distance from the Sun. Suppose, for example, that there were no greenhouse gases at all. The temperature would immediately drop to zero degrees Fahrenheit. The oceans would freeze, reflecting more sunlight and further cooling the Earth. (Some geologists think the Earth went through periods like that perhaps a billion years ago. It's called the "Snowball Earth" theory.)

On the other hand, suppose we succeed in increasing the greenhouse effect to 100%. What would the temperature of the Earth be then? We don't know exactly, but we do have a real example to look at. The surface of Venus should be somewhat warmer than the surface of the Earth, because Venus is closer to the Sun. However, that difference is not very large. It's possible that Venus could be very Earth-like. But we know it isn't. Venus has a poisonous atmosphere with a runaway greenhouse effect. When a Russian spacecraft sent a probe into the Venusian atmosphere, it recorded a surface temperature hotter than molten lead.

We don't know how big a perturbation it would take to tip the Earth's atmosphere into an entirely different state, one that might not be inhabitable by life at all. However, even the relatively small extant perturbations that the Earth has experienced can have dramatic effects. In various areas, the arctic permafrost has softened; low-lying islands have been inundated; and coastlines will change, to name only a few. We toy with atmospheric dynamics at our own peril. Some optimists believe that the sobering reality of Hubbert's peak will prevent us from doing irreparable damage. Yet this seems akin to hoping that a fatal heart attack will save a patient from cancer.

To be sure, the effects of the looming crisis could be greatly mitigated by taking steps to decrease the demand for oil. For example, with little sacrifice of convenience or comfort, we Americans could drive fuel-efficient hybrid cars. Such changes seem to be beginning, but there are powerful interests opposing them.

Before we turn to prospects for the future, a summation is in order. The followers of Hubbert may or may not be correct in their quantitative predictions of when the peak will occur. Regardless, they have taught us a crucial principle: the crisis will come not when we pump the last drop of oil, but rather when the rate at which oil can be pumped out of the ground starts to diminish. That means the crisis will come when we've used roughly half the oil that nature made for us. Thus, the problem is much closer than we had previously imagined. Beyond that, burning fossil fuels alters the atmosphere and could threaten the metastable state our planet is in. We have some very big problems to address.

So, what does the future hold? We can easily sketch out a worst-case scenario and a best-case scenario.

Worst case: After Hubbert's peak, all efforts to produce, distribute and consume alternative fuels fast enough to fill the gap between falling supplies and rising demand fail. Runaway inflation and worldwide economic depression leave many billions of people with no alternative but to burn coal in vast quantities for warmth, cooking and primitive industry. Dramatic change in the greenhouse effect results and eventually tips the Earth's climate into a new state hostile to life. Human civilization, not to mention all other forms of life in the biosphere, ends. In this instance, worst case really means worst case.

Best case: The worldwide disruptions that follow Hubbert's peak serve as a wake-up call. A methane-based economy is successful in bridging the gap temporarily, while nuclear power plants are built and the infrastructure for other alternative fuels is established. The world watches anxiously as each new Hubbert's peak estimate for uranium and oil shale makes front-page news.

Is there any hope for a truly sustainable long-term future civilization? The answer is yes. Stationary power plants can run on nuclear energy or sunlight. More difficult is a fuel for transportation. The fuel of the future is probably hydrogen. Not deuterium for thermonuclear fusion, but ordinary hydrogen to be burned as a fuel by old fashioned combustion, or to be used in hydrogen fuel cells that produce electricity directly. Burning it or using it in fuel cells puts into the atmosphere nothing but water vapor. Water vapor is a greenhouse gas to be sure, but unlike carbon dioxide, it cycles rapidly out of the atmosphere as rain or snow. Hydrogen is dangerous and difficult to handle and store, but so are gasoline and methane. Nature has not stored up a supply for us, but we can make it ourselves.

Of course you can't get something for nothing. Hydrogen is a high potential energy substance. That's precisely why it is valuable as a fuel. That energy has to come from somewhere. Where will we get the energy to make hydrogen?

Interestingly, one possible source is the potential energy stored in coal. There are extant industrial processes that combine coal and steam to make hydrogen and, inevitably, carbon dioxide. The process does not involve burning the coal. In principle, the carbon dioxide could be separated and stored ("sequestered" is the current buzzword). Where could it be stored? That little problem has not been solved yet. In any case the coal will eventually run out. We're trying to think long-term here.

Civilization as we know it evolved because there was a plentiful supply of oil in the ground, available for the taking. There is another cheap, plentiful supply of energy available for the taking, and this one won't run out for billions of years. It's called sunlight.

We now make very poor use of the sunlight that arrives at the Earth. Farmers use it to grow food and fibers for textiles. A little bit is collected indirectly in the form of hydroelectric and wind power. Here and there a few solar cells provide energy for one use or another. But by and large, it just gets absorbed by the Earth. It will wind up as heat in the Earth, eventually to be reradiated back out into space in any case, but we could learn to make better use of it along the way.

Sunlight is not very intense as energy sources go. The flux of energy from the Sun amounts to 343 watts per square meter at the top of the atmosphere, averaged over the entire surface of the Earth. By comparison, we Americans consume about a thousand watts of electric power each, all the time. Nevertheless, the solar power falling on the United States alone amounts to about ten thousand times as much electric power as even we profligate Americans consume.

Both sunlight and nuclear energy can be used to make hydrogen in a number of ways. There are chemicals and organisms that evolve hydrogen when sunlight is added. Electricity is made directly from sunlight in solar cells. Electricity can also be generated by using sunlight or nuclear energy as a source of heat to run a heat engine, such as a turbine, that can generate electricity. By means of electrolysis, electricity can make hydrogen from water. There is not much reason to doubt that hydrogen can serve as a fuel for transportation needs. At present, nuclear technology is far more advanced than solar for all of these purposes, but that could change in the future.

Technically and scientifically, the possibility and means exist to maintain and develop a civilization that provides everything we think we need, *without fossil fuels*. The future exists. The remaining question is, can we get there? And if it is possible to live without burning fossil fuels, why wait until the fuels are all burned up? Why not get to work on it right now, before we do possible irreparable damage to the climate of our planet?

Scientists are supposed to make predictions, and so I offer one. Civilization as we know it will come to an end some time in this century, when the fuel runs out. What that future looks like remains to be determined. This is different from normal scientific predictions in a crucial way. Usually, the scientist hopes that the prediction will prove to be correct, and merely making the prediction does not change the phenomenon in question. In this case I do hope the prediction will be wrong, and I hope that merely making the prediction will help obviate the problem.

Early life forms released oxygen into the atmosphere and buried carbon in the ground, preparing the planet for creatures like us. Now the planet is in our hands, and unlike early life, we are aware of our responsibilities and the possible consequences of our actions. What happens next is up to us.

Acknowledgements: I wish to acknowledge a number of colleagues who have read versions of this manuscript and made useful contributions to its accuracy: Dr. Robert Mackin, Dr. James Morgan, Dr. John Wettlaufer, Dr. Lynn Orr, Mr. Brad Haugaard, Mr. Mark Goodstein, Dr. Pierre Jungels, Dr. Tom Tombrello, Dr. Judith Goodstein.

REFERENCES

Bartlett, A.A. "Reflections on Sustainability, Population Growth and the Environment." *Population & Environment* 16, no. 1 (September 1994): 5 – 35.

Campbell, C.J. and J.H. Laherrère. "The End of Cheap Oil." *Scientific American* (March 1998).

Deffeys, K.S. *Hubbert's Peak: The Impending World Oil Shortage*. Princeton: Princeton University Press, 2001.

Duncan, R.C. "World Energy Production, Population Growth and the Road to the Olduvai Gorge." *Population and Environment* 22, no. 5 (May-June 2001).

Goodstein, David. *Out of Gas.* New York: W.W. Norton, 2004.

Hubbert, M. King. "Energy from Fossil Fuels." *Science* 109 (February 4, 1949): 103-109.

Ivanhoe, L.F. http://www.hubbertpeak.com/ivanhoe.

Lynch, M. http://sepwww.stanford.edu/sep/jon/world-oil.dir/lynch/worldoil.html
http://sepwww.stanford.edu/sep/jon/world-oil.dir/lynch2.html.

U.S. Department of Energy predictions.
http://www.eia.doe.gov/pub/oil_gas/petroleum/presentations/2000/long_term_supply/sld001.htm

U.S. Geological Survey, World Petroleum Assessment 2000: Description and Results.
http://www.usgs.gov/public/press/public_affairs/press_releases/pr1183m.html,
http://greenwood.cr.usgs.gov/energy/WorldEnergy/DDS-60/index.html#TOP.

Youngquist, Walter. *Geodestinies.* Portland: National Book Company, 1997.

INDEX